INVISIBLE RAINBOW

A PHYSICIST'S INTRODUCTION TO THE SCIENCE BEHIND CLASSICAL CHINESE MEDICINE

Changlin Zhang
with Jonathan Heaney

FOREWORD BY HARTMUT KAPTEINA

North Atlantic Books
Berkeley, California

Copyright © 2016 by Changlin Zhang. All rights reserved. No portion of this book, except for brief review, may be reproduced, stored in a retrieval system, or transmitted in any form or by any means—electronic, mechanical, photocopying, recording, or otherwise—without the written permission of the publisher. For information contact North Atlantic Books.

Published by
North Atlantic Books   Cover photos: (front) ritterdoron/iStock; (back) Nikola Nastasic/iStock
Berkeley, California   Cover and book design by Nicole Hayward

Figure 3.3 by United States Air Force Senior Airman Joshua Strang (public domain)
Figure 13.2 by Hans Nilsson, Kungliga Opera, used under Creative Commons license (http://creativecommons.org/licenses/by-sa/3.0)
Figure 13.6 by Paata Vardanashvili from Tbilsi, Georgia (Swan Lake) and used under Creative Commons license (http://creativecommons.org/licenses/by/2.0)

Printed in the United States of America

Invisible Rainbow: A Physicist's Introduction to the Science behind Classical Chinese Medicine is sponsored and published by the Society for the Study of Native Arts and Sciences (dba North Atlantic Books), an educational nonprofit based in Berkeley, California, that collaborates with partners to develop cross-cultural perspectives, nurture holistic views of art, science, the humanities, and healing, and seed personal and global transformation by publishing work on the relationship of body, spirit, and nature.

North Atlantic Books' publications are available through most bookstores. For further information, visit our website at www.northatlanticbooks.com or call 800-733-3000.

Library of Congress Cataloging-in-Publication Data

Zhang, Changlin, author. | Heaney, Jonathan, author.
  Invisible rainbow : a physicist's introduction to the science behind classical Chinese medicine / Changlin Zhang, with Jonathan Heaney.
LCCN 2015049047 (print) | LCCN 2016003076 (ebook) | ISBN 9781623170103 (print)
  ISBN 1623170109 (print) | ISBN 9781623170110 (ebook)
LCSH: Medicine, Chinese. | Acupuncture. | Electromagnetism—Physiological effect.
LCC R601 .Z6852 2016 (print) | LCC R601 (ebook) | DDC 610.951—dc23
LC record available at http://lccn.loc.gov/2015049047

1 2 3 4 5 6 7 8 Sheridan 21 20 19 18 17 16
Printed on recycled paper

Praise for *Invisible Rainbow*

"In *Invisible Rainbow*, Changlin Zhang not only opens the gates to a scientific understanding of Classical Chinese Medicine, but he also opens doors to break out of the prison of naturalistic paradigms of our Western world, paradigms that reduce man to human capital, human material, and biochemical machine. Medically, the human material has to be kept functioning as long as possible to secure maximum medical and economic profit. As a biophysicist, Zhang opens our eyes to all the invisible, and our ears to all the inaudible reality that surrounds us, which we like to exclude from our views in order to keep our world as easy to control and grasp as we like it."

—HANS-JOACHIM HAHN, FOUNDER
AND ORGANIZER OF PROFESSORENFORUM,
AN ORGANIZATION OF GERMAN PROFESSORS

"Many people will enjoy reading this book and benefit from it in various aspects."

—RUIXIANG CHEN, PROFESSOR AT
BEIJING UNIVERSITY OF CHINESE MEDICINE

"The author, Professor Changlin Zhange, describes the acu-points and acu-meridians through a physics lens, with vivid language, proficient knowledge, and profound insight into modern physics."

—GUOZHEN CHEN, PROFESSOR OF
PHYSICS, DONGWU UNIVERSITY, TAIWAN

*To the One who wrote the book with infinite pages
and let me read a few of them.*

ACKNOWLEDGMENTS

Max Reinhard in Germany, who sponsored the research project for more than fifteen years with incredible patience, tolerance, and confidence.

Hartmut Kapteina, professor of music and music therapy. My closest research partner and dear friend. He also wrote the beautiful foreword for this book.

Hans-Joachim Hahn, who is dedicated to rebuilding the soul and the spirit of universities in Europe and contributed greatly to getting the German version of this book published.

Jian-Ping Chu in Switzerland, Bisong Guo in Scotland, and Dan Jiang in England—doctors of Chinese medicine, who gave me a lot of help with discussions and encouragement.

Valerie Luff, a nurse who spent her career in missionary hospitals in Nepal, India, and Bangladesh. I met her when I was a visiting scholar in the University College London. She is my godmother and my real spiritual mother; particularly as I lost my own physical and spiritual mother in China when I was a teenager.

Ding-Zhong Li, professor of Chinese medicine in Beijing. A pioneer of modern scientific research into acupuncture. In 1996, he nominated me to be the leading scientist of the national project on meridian research.

Ke-Hsueh Li, one of a few theoretical physicists who truly, deeply understood the profound background of modern physics. My teacher during postgraduate studies at the Chinese Academy of Sciences, Beijing, from 1978 to 1981, and my research supervisor in Germany from 1991 to 1994. The first person to hear my idea of electromagnetic standing waves inside human bodies, many important experiments in this book were designed in accordance with his advice.

Elisabeth Reschat, Dietrich Stoeckl, Thomas Lastring for their work translating, refining, and improving the German version of this book as well as obtaining the copyright permissions for a number of figures in this book.

Hans-Juergen Stoeckmann, a professor at Marburg University, for kindly checking the statements from the viewpoint of physics in the German version.

Peter Weinberg, a professor at Hamburg University, gave many constructive suggestions for this book, and his doctoral student Christof Ziaja, who found a new statistical distribution with our coherence meter (or biometer) in their sleep lab.

Xiang-Yang Zhao in Germany, a brilliant young scholar and physicist who assisted the Chinese translation by providing many critical and constructive opinions.

Zhong-Shen Liu, my gifted, old and dear friend at the Heilongjiang University of Traditional Chinese Medicine, for further refining the Chinese version.

My beloved family, my wife, Xinger (Angela), and our son, Zhiliang (Albert). Our son was the first person tested with music measurement and, like his mother, always supports me with love, hope, and persistence.

My students, Zhi-Hao Zheng, Min Zhang, Hong-Zhen He, Yi-Bin Pan, Hai-Ou Wang, Bo-Lin Xu, Yuan-Dong Zhou, Li-Jun Meng, and many others in China, who did so much important work along with my former assistant Da-Lin Li, who helped to edit and publish *Current Development of Biophysics* at Hangzhou University Press in China in 1996.

Jonathan Heaney, who spent countless hours improving and rewriting the entire English manuscript sentence by sentence and word by word. Jonathan would like to thank his wife, Mei Ling Chang, for her constant support, encouragement, and feedback as well as Cindy Colson and Jillian Heaney for their advice and feedback.

Actually, this list is endless; it is impossible to acknowledge all of those who helped me in any book with limited pages.

CONTENTS

Foreword / xi
Preface / xv
Introduction / 1

PART 1: THE WORLD OF THE BLIND / 5
1. Revisiting "Blind Men Study an Elephant" / 7
2. Spiritualized Physics and Materialized Psychology and Biology / 14
3. Inaudible Music and the Invisible Rainbow around Us / 28

PART 2: TWO PARADIGM SHIFTS—ONE IN MEDICINE, ANOTHER IN SCIENCE / 41
4. Major Changes in the Medical Market / 43
5. Queen Victoria Studies TV / 54
6. Blind Scientists Discover the Rainbow / 70

PART 3: DEVELOPING THE CONCEPT OF STRUCTURE / 85
7. A New Continent in Science: The Dissipative Structure / 87
8. Standing Waves and Wave Superposition / 95
9. Wireless Communication inside a Body / 104

PART 4: FIELD AND WAVE ASPECTS OF BIOLOGY / 117
10. Powerful Resonance: A Secret Means of Transferring Energy and Information / 119
11. The Mysterious Aura: From Religious to Practical / 133

PART 5: MEASURING COHERENCE / 145
   12. Facing Complex Systems: The End of Reductionism / 147
   13. How Much Beauty Is There in a Ballet? / 171
   14. Measuring the Invisible Rainbow / 182
   15. Coherence in Medicine and Health Care / 194

Epilogue: Consciousness, Spirit, and Conscience in Science / 201
Afterword: Endless Exploration / 208
Notes / 211
Bibliography / 216
Index / 219
About the Authors / 227

FOREWORD

In his book *Nada Brahma: The World Is Sound,* Joachim-Ernst Berendt, the famous jazz researcher and cross-thinker in musical science, says that Japanese Zen masters asked their students to do the following meditation exercise: "If you delete purpose and sound, what do you hear?"[1] In this exercise, the meditating person practices the art of excluding sounds entering from outside, and the ability to listen to what is inside. In this practice, with every step in which the adept succeeds in actually not hearing anything anymore, he or she enters into a new space of hearing, at first being filled with all the sounds of the body, like breathing, blood circulation, and the friction of muscles and bones. Withdrawing perception even further from this space of hearing, the adept will proceed to different inner worlds of sound, finally entering the innermost sanctuary of hearing, the vibration of atoms and molecules: "a bright, silver ringing." Alfred A. Tomatis, the famous French ear, nose, and throat specialist, who did research on the secrets of hearing throughout his life, called this phenomenon "the sound of life."[2] Yes, the ear was actually capable of hearing the vibrations of elementary parts, because the cilia, the antennae through which the hearing cells receive information, have a radius the same size as a molecule.

An Indian healer was invited to a congress of psychologists to report about her therapeutic work. In her talk, she said,

> If I wanted to illustrate to you my healing techniques in the words and terms of our culture, I would talk about ancestors, demons, spirits, and extrasensory powers, and you would push it aside as some superstitious nonsense. Therefore, I want to describe my therapy to you in your terms, designed by traditional European and American natural science and medicine.

She explained the theoretical foundations of her treatment:

As you all know, matter consists of elementary particles. Each of these particles vibrates at a certain frequency. Just imagine if you were able to hear these vibrations. Each vibration would have only one tone.

The individual atoms are chemically connected to molecules; viewed in terms of music, this is a chord. The molecules are built into cells; musically, these are larger combinations of chords. The cells form most parts of the body: bones, tissues, muscles, organs; together they create a symphony of high complexity, playing wonderful music. Human spirit, emotions, and desires shape the forces, which maintain this body music in a melodious and harmonic interplay. If a person gets sick, this interplay is disturbed, and body music sounds inharmonious and dissonant. In my culture, the healer has learned to capture the sound of body organs by way of meditation and concentrated perception, and to change it through appropriate procedures.

In this process, the healer is aware of the makeup of his own body, which he experiences in the same way as a vibrating and sound-producing orchestra. When he perceives the illness of the patient, he also experiences the pain and disturbance like the dissonant sound of a garbled piece of music. He has learned the correct remedy for healing the patient, and how disturbed harmony in the organism can be restored—through presenting a healing plant, a certain song, or a healing dance, or perhaps a change in life circumstances or the resolution of a social conflict.

This book by Changlin Zhang, professor of biophysics at Hangzhou University and the University of Siegen, about the invisible rainbow, puts an end to the time when the stories of Indian healers and Zen monks can be dismissed as esoteric. He is a proven insider of the treasure of experience from thousands of years of Eastern healing traditions, and at the same time is a highly qualified expert in Western natural sciences like physics, chemistry, and biology. With appropriate skepticism, he describes the many research experiments with which natural scientists have tried to explain apparently inexplicable phenomena in the areas of acupuncture and homeopathy.

In ways similar to the Indian healer, he leads the reader step-by-step from the simplest physical data into increasingly more complex areas of human life and existence. By combining the results of modern physics, chemistry, and biology with the experiential knowledge of Chinese and Indian healing arts, a new holistic picture of the world and humanity develops. In it, acupuncture and homeopathy appear as treatments well established by natural science.

A colleague recently told me about a lively discussion at a psychology seminar about an old philosophical question: What is the true origin of human existence, spirit or body? While listening to her present the individual arguments that were presented for one side or the other, I felt these familiar viewpoints touch me in a new and unprecedented way. I mentioned that I had recently read the manuscript for Zhang's *Invisible Rainbow,* and since then, human existence appeared to me in a completely changed light; this is what her presentation of the arguments about body and spirit had caused me to become aware of.

If all matter, and in the same way also the body, is nothing but vibrations of the smallest elementary particles interwoven with each other in complex and manifold ways, and the motions of our thinking, feeling, and acting can be understood as variations and especially modified expressions of such vibrations, then body and spirit, body and soul, immanence and transcendence, living and dying, death and resurrection, and many other elements that make up our lives must no longer be conceived as dichotomous. Rather, they represent different stages and forms of one and the same vibrating medium, which keeps changing in seamless transitions into constantly new dissipative and stable patterns of waves.

Zhang starts this book with the foundations of modern natural science, then proceeds from insight to insight and portrays impressive scientific and technological progress. At the same time, he points out the manifold forms of suffering, desolation, and loss of purpose connected with this progress. He perceives a central cause for these developments in the fact that decisive insights in modern physics about space, time, and matter are not being recognized by the applied natural sciences, especially medicine, and the humanities, especially psychology and sociology.

Instead, he discovered that this groundbreaking progress in physics was already circulating as anticipative knowledge in ancient high cultures. Even in European culture we find such context when we read in the first pages of the Bible, that in the beginning God created heaven—light—perhaps the original vibration, caused by the Big Bang, that determines the whole cosmos with its organizing and energizing effect, for us the invisible rainbow. Even before the creation of light, the Bible says, "And God said ..."; was this act of speaking the Big Bang, which, up to now, keeps the whole cosmos vibrating, or was it the dance of Shiva in Indian mythology, or Nada Brahma in the Buddhist tradition?

The creation of sound as the first creative act permeates the creation myths of many cultures.[3] This original sound and original light are the inaudible music and the invisible rainbow, and groundbreaking insight about them can be gained from this book.

A final note: this book does not stop at the level of foundational research and general theory. It demonstrates how new methods of diagnosis and therapy can arise through the context presented. A practical example happening right now, in the practical completion and field testing phase, is holistic measuring to conceive the total psychophysical state of living organisms.

HARTMUT KAPTEINA
UNIVERSITY OF SIEGEN

PREFACE

Isaac Newton once said, "to myself I seem to have been only like a boy playing on the seashore, and diverting myself in now and then finding a smoother pebble or a prettier shell than ordinary, whilst the great ocean of truth lay all undiscovered before me."[1] In this context, I consider myself fortunate to have been granted the opportunity to find a pretty shell or two to appreciate and explore. More precisely, I am profoundly grateful to the ocean's creator, who granted me the opportunity to gaze at and appreciate the breathtaking beautiful rainbow inside and around our bodies, to hear the charming symphony of electromagnetic waves in our bodies. Furthermore, I feel blessed to have learned to measure and calculate the degree of harmony in the color of the rainbow and the melody of the symphony in Hilbert space, that is, infinite-dimensional space.

The motivation behind writing this book is to overcome the language barrier between those involved in biological and medical science and physicists. Originally, the challenge of venturing into this new area arose from attempting to address two questions: first, could a scientific explanation for the mechanism of acupuncture and many other ancient and holistic medicines be found? If this was indeed possible, the second question was how best to pursue this goal. The answer to the first question cannot be found within the current understanding of Western medicine. Instead, one must venture to the frontier areas of modern physics.

The cross-disciplinary nature of the inquiry presents a serious communication problem for those accustomed to Western medicine and conventional biology. Encouraged by many of my friends in the medical field, I decided to write a book to bridge this communication gap. In doing so, I employed descriptions like "invisible rainbow" to provide a conceptual familiar passage to explore concepts like the spectrum of electromagnetic waves, standing waves, dissipative structures, interference patterns of standing waves, and resonance cavities. Similarly "inaudible music" provides an intuitive path to approach ideas like statistical distribution and the resonance effect. Finally, posing the question "How beautiful is a ballet?" provides a way to convey the essence of what the application of combinatorial mathematics, general distance, and Hilbert space are trying to achieve.

The core intent of this book is to provide a brief description of the invisible rainbow and inaudible music in our bodies. They exist beyond our senses, but in a sense you will be able to "see" and "hear" them yourself in the way that physicists are able to "see" the structure of atoms through experimental results and rational reasoning.

The book has been divided into parts, the focus of which, as outlined in the introduction, ranges from exploring the nature and history of scientific research, to explaining the background theories in physics, to conveying the experimental evidence and mathematical analysis of the research, and finally to considering the implications for health care moving forward. My intent in doing this is to allow readers to select the parts to read that offer them the most given their current level of knowledge and their objectives.

As with all scientific advances, so much is owed to the great scientists whose achievements we seek to build upon. In my field I am profoundly grateful to the pioneers of scientific research in this area in recent centuries. These include many German, British, Austrian, Belgian, American, and Greek scientists. I did my best to acknowledge their contributions in this book. Of course, there are many more than those I have mentioned, and it is ultimately impossible to acknowledge all of them.

I am also grateful for the opportunity that writing this book offers me to introduce some important research performed by Chinese scientists in the 1970s and 1980s, when China was mostly isolated from the outside world. Almost all of their important work was published only in Chinese and is still largely unknown in the rest of the world. They were the direct pioneers of modern science's exploration into the secrets of the ancient practice of acupuncture, and much of their work is still valuable today. It is my hope that scientists in the West will be motivated to explore, reproduce, and verify or possibly even dispute their findings in order to build on the work of these pioneers.

INTRODUCTION

From their ancient roots, traditional therapies have enjoyed a revival in the recent decades. Their burgeoning popularity aside, they are still generally viewed unfavorably by Western medicine and the mainstream biological sciences. This attitude ultimately stems from the difficulty of explaining how they work within a scientific framework. Central to this difficulty is the predisposition of biology and medicine to focus on reductionism and the nineteenth-century concept of materialism. While this focus enabled the formidable achievements of Western medicine in combatting acute diseases and developing surgical techniques, it came at the cost of holistic understanding.

While this book deals with a number of traditional therapies, it focuses on Eastern therapies, in particular Classical Chinese Medicine and acupuncture. Classical Chinese Medicine (CCM) differs from Traditional Chinese Medicine (TCM) in that CCM refers to a system of medicine as it was practiced prior to the Cultural Revolution of the 1960s and 1970s. TCM, whose Chinese name is more accurately translated as "systemized Chinese medicine," is the product of attempts by the government of the Cultural Revolution era to standardize the knowledge for mass education and to integrate a Western-medicine style of thinking into Chinese medical education. In doing so, much of the rich knowledge and understanding of CCM was lost or distorted.

Advances in physics in the last century have continued to move progressively farther from materialism into the realms of waves and fields and from considering closed isolated systems to open ones. It is these advances that have enabled a scientific framework to be developed that correlates with the concepts of Classical Chinese Medicine and explains how acupuncture functions.

Ultimately these advances in physics enabled the discovery of the electromagnetic body. To put this discovery in context, it is helpful to investigate the world that exists beyond the scope of our senses and also to appreciate the role that these sensory limitations play in how we communally frame what constitutes reality. As such, part 1 of this book explores the idea of how we can become aware of things that exist beyond the range of our senses.

Chapter 1 explores a thought experiment about how a world populated by blind people would go about discovering light. Would these people ever be able to appreciate the nature of a world that is so readily apparent to us? How would their scientists go about discerning this knowledge?

Many parallels can be made between the supposed journey of the blind scientists in chapter 1 and our recognition of the mostly invisible world of electromagnetic waves and fields in the twentieth century. Chapter 2 explores how physics progressed from nineteenth-century materialism to energy, fields, and the enigmatic realm of quantum mechanics. Interestingly, while this movement was occurring in physics, biology and medicine focused their development more toward the atoms and molecules of nineteenth-century materialism.

As the key for scientific understanding of acupuncture lies in the realm of electromagnetic waves and fields, chapter 3 explores what scientific discoveries of the last century tell us about them. This world that mostly exists beyond the range of our senses is extremely rich and dynamic and is in some ways akin to an invisible rainbow and inaudible music. What would it look like, with its invisible structures of interference patterns, if we could perceive it directly?

Part 2 of this book looks at the paradigm shifts that have occurred in science and medicine. In doing so, it also reviews the path of scientific research into acupuncture. After being all but abandoned for much of the nineteenth and twentieth centuries, traditional therapies have enjoyed a significant revival in the past few decades. Chapter 4 discusses the causes behind this, while also explaining the reasons for the original decline of these traditional remedies in the face of Western medicine's ascendancy.

Chapter 5 looks at science's earlier struggles in discerning the underlying model or structure to support the evident function of acupuncture. This moves from the early failed attempts to discern an anatomical basis for the meridians to the extensive studies in China, which provided strong statistical support for a genuine effect, and how their results correlate with the historical texts. Finally, it discusses the later hypotheses that consider wave and field phenomena and their importance in regulating our bodies, which proved to be closer to the truth.

The discussion of the scientific exploration of acupuncture continues in chapter 6, where questions as to the nature of acupuncture are posed in light of more recent electronic scientific investigations. This leads to further insights into what the skin resistance measurements that typified this investigation were actually measuring: the strength of fields or energy distribution within the body. This discovery is key to understanding many of the puzzling phenomena associated with acupuncture.

Part 3 investigates the concept of nonstatic dissipative structures and the wave phenomena associated with them. This leads to an understanding of how acupuncture can influence these structures and their importance to the body. Scientific recognition of dissipative structures occurred incrementally. Chapter 7 explores the developments that led to this discovery, starting with the failed quest for perpetual motion and the concept of entropy. This is followed by an exploration of science's subsequent movement from considering closed systems to open systems, equilibrium states and far from equilibrium states, and static structures to dissipative structures.

In order to explain how inserting needles at certain points on the body can influence the health of a patient, chapter 8 explores the classical physics of waves, including wave superposition, standing waves, and the importance of boundary conditions. This serves to illustrate the mechanism through which acupuncture can influence the energy distribution in the body.

The body's chemical messenger system (hormones) and electrical messenger system (the nervous system) are well known. Chapter 9 explores the infinitely more complex electromagnetic messenger system as an additional method by which the body coordinates and controls its multitude of components. This lays the framework for the concept of the electromagnetic body, which represents the next step for biological and medical investigation.

In order to explain the mechanisms of the electromagnetic body and the phenomena associated with it, part 4 expands on the wave and field aspects of biology. Chapter 10 discusses the phenomenon of resonance and its ability to transmit energy and information throughout the body. In addition to explaining the role of the electromagnetic body in health maintenance, it also facilitates an explanation for the health implications and the importance of belief.

In addition to providing an explanatory mechanism for acupuncture, the electromagnetic body also provides a scientific rationale for the aura. Chapter 11 deals with how it can be interpreted in light of the electromagnetic body and how scientific measurements and known wave phenomena correlate with an aura that exists at the fringe of our visual senses. The electromagnetic body is constructed through the interaction of a practically infinite number of electromagnetic waves. The concept of coherence is central to assessing the nature of this interaction. Part 5 explores the measurement and scientific analysis of coherence.

The reductionist approach to correcting an orchestra would be to subdivide it into progressively smaller subparts and ensure that these are functioning correctly. This approach overlooks the importance of the cooperation and coordination of the instruments in forming a symphony. Chapter 12 expands on the concept of coherence and explores the contrasts between the underlying linear thinking in Western thought, with its roots in the Egyptian and Greek civilizations, and the more holistic thinking of the Eastern civilizations of China and India. It then reflects on the current collision occurring between these cultures and what their integration can bring to science.

Chapter 13 further explores the idea of coherence and its importance in natural medicine and the electromagnetic body. Developing an objective means of assessing this is critical to advancing our scientific understanding of the electromagnetic body. Chapter 13 uses everyday language to explain the basic mathematical idea that serves as the foundation to achieve this objective.

This mathematical idea is then expanded in chapter 14 to explain how the coherence, and therefore the state, of the electromagnetic body can be measured. It is through the statistical methods discussed in this chapter that the condition of an individual's electromagnetic body can be presented in a readily understandable manner.

Finally, chapter 15 looks at the implications of these developments for health care. In particular it looks at how a new approach to health care needs to consider the body's state in relation to the rhythms of our everyday life. The chapter also explores its potential for understanding and enhancing our use of alternative therapies, particularly in treating chronic conditions, and for monitoring our well-being.

PART 1

THE WORLD OF THE BLIND

CHAPTER 1

REVISITING "BLIND MEN STUDY AN ELEPHANT"

One day, a foreigner presented an elephant to a king. The elephant, an exotic animal in this kingdom, had never been seen before. The king called several blind men to his palace and asked them what the elephant looked like after they touched it. The blind man who had touched the leg said: "Oh, Your Majesty, it is like a pillar." Another blind man, who had touched the ear, said: "Oh, Your Majesty, it is like a big leaf." The king and his ministers laughed at their answers.

The situation in basic scientific research is sometimes quite like this famous tale, or even worse. To clarify, basic scientific research refers to fundamental exploration to pursue the knowledge of reality and truth, as opposed to applied science and technology. Returning to the tale, it is misguided to dismiss the blind men. They were serious, intelligent, and selflessly honest. Given their experiences and hypotheses, their explanations were as accurate as they could be. Exactly the same can be said for the earnest labors of countless scientists throughout history and for those operating today in any area of basic scientific research.

To elaborate, let us revisit the tale from a slightly different viewpoint. The intent is to provide an insight into the realities of basic scientific research, thereby conveying a sense of its inherent difficulties. Furthermore, this tale also serves to illustrate why these challenges are particularly prominent for research probing the frontiers of scientific understanding.

A New Version of "Blind Men Study an Elephant"

Let us first suppose a world where all people were blind, as were all of their ancestors. This world's level of intellectual advancement is similar to ours.

As in our world, there are many gifted scientists who tirelessly pursue truth and knowledge.

Meanwhile, let us suppose that the elephant in their world is significantly larger than those of our world, say as big as a large building, or even a hill. At this world's first encounter with an elephant, it naturally became the focal point of contemporary scientific research.

Let us suppose that three outstanding research groups, whose leaders were Professor A, Professor B, and Professor C, spearheaded this scientific endeavor. Being specialists in different fields of professional knowledge, they studied the same elephant from different perspectives. Consequently, they proposed different models, hypotheses, interpretations, and theories about the nature of this alien manifestation.

The first group, led by Professor A, started its research from the tusk of the elephant. They made careful and precise measurements, performed sophisticated calculations, undertook highly philosophical and rational data analysis, and eventually proposed the subsequently famous Carrot Hypothesis, Carrot Model, or Carrot Theory. That is, they thought the nature of the elephant was akin to a huge carrot. Consequently, they predicted the existence of big leaves on the top of the big carrot. It is possible to imagine their excitement when they touched the ears of the elephant, firmly believing them to be the leaves of the carrot. This development served as a powerful verification of their prediction and provided compelling evidence in support of their brilliant theory. If this world possessed a Nobel Prize, Professor A's group might be likely candidates to receive the laurels.

Around the same time, Professor B's group commenced its investigation, starting from a leg of the elephant. Their analysis led them to conclude that the shape of the elephant was similar to the trunk of a large tree. They proposed the Tree Model or Tree Theory. However, they soon encountered another leg. Accordingly, they adapted their theory from the Tree Model to the Forest Model in accordance with these new experimental results. As with the Carrot Theory, the verification supplied by the discovery of the elephant's third and fourth legs generated much excitement.

Professor C was the brilliant former protégé of Professor B, and the team further developed Professor B's research. After careful and precise measurements of the four trunk-like pillars and their surroundings, they found that there were only four pillars—not more. Also, these four pillars were somehow arranged according to the four corners of a rectangle. Furthermore, they discovered that the temperature within this rectangle was consistently lower than that of the surrounding areas. Consequently, between the four pillars was something that isolated the region from the heat of the sky (they did not refer to sun in the sky because they had never seen it). In accordance with this new evidence, they proposed Table Theory as an improvement and development of Forest Theory.

In order to verify the new Table Theory, they constructed a technologically advanced instrument—an extraordinarily long ladder. It is easy to envision their sense of elation upon reaching the underside of the elephant's torso. From their perspective they had discovered the huge table, thereby obtaining compelling evidence verifying the prediction of Table Theory.

Naturally, as their world's science continued to develop, they would discover that all these theories were only "special theories," valid within special areas and situations. Finally, after many academic conferences, seminars, discussions, controversies, disputes, possibly even quarrels and conflicts, they might, if they were lucky, establish a "general theory" or "unified theory." This would include all the existing special theories that were supported with reproducible experimental evidence. At this point, they might be able to comprehend what the elephant looked like, in a manner consistent with what we would perceive at first glance.

Eventually, once this ultimate authorized unified theory was well established in their education system, coming generations in their world would no longer be puzzled about the appearance of the elephant. They could dutifully study the theory and subsequently impart the knowledge to others in the manner of an authority on science.

However, it is useful to expand on the meaning of the statement "if they were lucky." In the case of the story, it means if the elephant was not too big. Consider the situation if the elephant was as big as our earth, our solar system, or our galaxy. What would the future hold for these scientists? How could they approach a unified and final theory?

This point serves to illustrate how misguided it is to try to establish a final theory for an infinite object, such as our universe, or even to establish a unified and final theory of physics.

Blind People Study a Rainbow

In many cases the situation facing basic scientific research is even more challenging than that confronting the blind scientists studying an elephant. In fact, the challenge confronting this avenue of research is more akin to "Blind People Study a Rainbow" and "Deaf People Study Music." In addition to being invisible and inaudible, the object of our research is dynamic.

To illustrate, let us continue our thought experiment in the world of the blind. Moving forward from their elephant conundrum, let us explore how their scientists would become aware of and study the colorful world around them. Would their scientific development ever extend to discovering the invisible and untouchable rainbow?

This experiment has supposed a world where every person is blind, as were all preceding generations of people. Therefore, in this world, there are no such words as *red, orange, yellow, green, blue,* and *purple* to describe colors. There are not even words like *light, bright, dark,* or other words relating to vision. This world is home to many highly capable scientists. Given this, let us present a couple of questions for discussion.

The first question arises in comparison to a property of our everyday world that is abundantly clear to us. "Is it possible for their scientists to discover that their world is colorful?" If this is possible, the next question should be, "How would these scientists go about describing the beauty of this discovery to the blind inhabitants of their world?" How does one achieve this without the existence of words related to color and light? This is the greatest challenge in writing this book.

In my opinion, the blind people would eventually discern that their world is not completely dark, but full of light and beautiful colors. An analogy can be drawn to the way that we now comprehend the existence of imperceptible ultraviolet rays, microwaves, and radio waves. The seemingly simple task of appreciating their colorful world would pose an extreme challenge. Only the most eminent philosophers and scientists, building on many years of exceedingly hard work and abstract reasoning, could approximately approach it.

Their journey of discovery might begin with the observation that the temperature on the south wall is usually higher than that on the north wall (assuming they were living in the northern hemisphere of their world). Similarly, it might occur to them that in the morning, when they notice their environment regularly gets warmer, the east-facing wall is usually warmer than the one facing west. Correspondingly, the arrival of the afternoon, when temperatures regularly declined, would be associated with exactly the opposite phenomenon.

They would also find that these temperature discrepancies exhibited some correlation with the weather. The disparity would disappear whenever it was raining. However, the variation would not always accompany the absence of rain: the inhabitants of this world would not be able to tell whether the sky was clear or cloudy. Furthermore, whenever this difference occurred, they would also feel the temperature contrast on their skin. It would come to their attention that all these effects existed only outdoors. Evidently, when indoors, the roof and walls could effectively block some unknown factor or sensation propagation. This could also be described as "Qi," "vital energy," or any other term their people chose to use. (Qi is central to Classical Chinese Medicine and martial arts; its meaning can be approximated by the idea of "life force" or "energy.")

In addition to subjective sensations and objective temperature measurements, careful observations of the behavior of animals and plants would lead them to infer that something existed that was unknown to them. Each passing year would foster a growing awareness of "something" existing in their world that was beyond both the scope of the senses and the range of their language.

Their scientists, like ours, would prefer objective instrumental measurements to subjective feelings. This inclination would lead to the development of many precise instruments to quantitatively detect and measure this unknown factor. Subsequently, other scientists, specializing in theoretical work, would delve into the freshly accumulated experimental data and experience and propose various models, hypotheses, theories, and mathematical formulas. Further experiments would then be designed and conducted to test these theoretical proposals with the results, verifying some and refuting others. Consequently, some theories would be discarded, some improved, and some would be merged into more general theories.

Depending on their rate of progress, this process of scientific exploration would continue for dozens, hundreds, or even thousands of years. Finally, after much thought, discussion, and controversy, their scientists would finally recognize the existence of the electromagnetic wave.[1] At this point, they would have developed instruments that enabled precise, large-scale measurement of these waves. A multitude of subsequent measurements in many different places would provide a wealth of data about the strength of the various electromagnetic waves according to their wavelengths. Finally, after extensive complex mathematical analysis, scientists would realize that their world was very "colorful."

This discovery might inspire the creation of a cumbersome textbook abounding with complex experimental data and esoteric mathematical formulas. Consequently, only the most devoted physics students would be able to learn and comprehend this knowledge. Some scientists, having gained an appreciation of this beautiful realm, would feel obliged to attempt to write popular science books to convey a sense of it to the public. This would be challenging, as they would have to avoid mathematical language and minimize terminology while maintaining an engaging style. However, their greatest challenge would be describing the colorful world without access to language relating to color and light.

To achieve this, they might employ metaphors to help the blind layperson imagine, conceive, and to some extent even perceive this unseen world. They would also have to invent special new words, like *atom, electron,* and *quark* in our world, to describe these imperceptible phenomena. If they happened to invent the word *light* to describe the electromagnetic waves between 400 and 700 nanometers[2] and used the same words we use for colors to describe the appropriate wavelengths of visible light,[3] their recognition of the colorful world would begin to resemble an exceedingly simplified abstraction of our experience.

Evidently, this is a highly contrived scenario. However, we must face the fact that we are all blind people to most electromagnetic waves—those beyond the very narrow range of 400 to 700 nanometers. We must also concede that we are all deaf people to most mechanical (sound) waves beyond the narrow range of 20 to 20,000 hertz. In fact, the story of how physicists in our world came to recognize the invisible atom and the invisible radio wave bears remarkable resemblance to the stories of "Blind People Study a Rainbow" and "Deaf People Study Music."

Blind People Study a Clever Mouse

Aside from the challenge posed by language, we also struggle with many other limitations, such as our sensory systems and thinking ability. To illustrate how scientific research is inhibited by these and other often unconsidered limitations, let us continue developing the thought experiment in the world of blind people.

Suppose a blind person, by chance, came across a mouse. As the nimble, highly intelligent mice that inhabit this world had never been detected before, it was a somewhat traumatizing experience. Naturally, he would share the experience and attempt to convey the horrible sensation to his friends. He would also be quite eager to present the novel animal to them. However, the likelihood of being able to do so would be exceedingly small. In terms of modern science, the reproducibility of the experiment is very unlikely.

If sharing his experience ensued as expected—without the presentation of any mice—how would his friends react? At best, they might conclude that he had confused a nightmare with reality. Some of them might even question their friend's honesty and accuse him of playing a joke or even telling tall tales. It is probable that the unfortunate person would be unable to show any scientific evidence to verify the claim.

He might persevere until one day he invented some new technology capable of catching the mouse with certainty. Considering the high intelligence of the mice, this difficult task might take decades or even centuries. Perhaps fate would be slightly kinder. Other blind people might have experienced a similar startling event and proceeded to tell similar stories. However, no one would be able to produce any objective experimental evidence. The accumulated stories might attract the attention of some scientists who were open to new phenomena, leading them to consider the possibility that these stories alluded to something beyond current scientific understanding. A few would risk their reputations, and possibly even their jobs, by investing in daring, unsuccessful attempts to catch the animal. As this exploration proceeded, the objective existence of the alien animal, namely the mouse, would become an unresolved question in the world of blind people.

After decades or even centuries of exploration, including many careers blighted by failure as well as some successful endeavors, the technology for catching mice would be successfully developed. Possibly representing a relative level of technology akin to the huge particle accelerators in California and Geneva, this equipment would allow anyone in their world to touch mice at will. In other words, the experiment would be objective, have excellent reproducibility, and be a science success story. The experiment's success would oblige the mainstream population to accept the existence of mice. At this point, the existence of mice would progress from being a fantasy to being a topic of conventional study.

This story serves to illustrate that the acceptance of something in the scientific community is not necessarily based on reality; instead, it is a function of belief. More precisely, scientific acceptance is based on the belief of the majority and is thus ascribed with all the inherent advantages and disadvantages of a democratic system.

CHAPTER 2

SPIRITUALIZED PHYSICS AND MATERIALIZED PSYCHOLOGY AND BIOLOGY

> While psychology has been moving toward the mechanical concepts of nineteenth-century physics, physics itself has moved in just the opposite direction.
> —HENRY P. STAPP, *Mind, Matter, and Quantum Mechanics*

If we move on from the preceding chapter's discussion of the challenges inherent in investigating phenomena that exist beyond the senses and instead consider the dynamics of scientific understanding in our world, it becomes apparent that the conceptual framework employed by mainstream society generally lags far behind developments in contemporary science. This prevalence of outmoded concepts even permeates the scientific community, where the concepts of the majority of scientists, particularly outside their area of specialization, harbor superseded conceptual models.

As the opening quotation of this chapter brings the concepts of nineteenth-century physics to the foreground, it is helpful to elaborate on their nature. At that time, the concepts of physics were dominated by the philosophy of materialism. This holds that the essence of the world is material and as such is visible, has definite weight, occupies definite space, and is composed of miniscule solid particles. Due to the remarkable success of physics in explaining numerous phenomena during the nineteenth century, its concepts became revered. Materialism also provided the basis for the philosophy of Marxism, which dominated half the world in the twentieth century. Even today, material considerations still exert the utmost influence on humanity's collective consciousness.

The enduring influence of the concepts of nineteenth-century physics is evident in the numerous materialism-biased misperceptions present in the modern mind-set, including that of the majority of scientists. For example, most people believe that science views the constituents of reality as always being solid and materialized. These include the impressive, and intuitively reasonable, ball-and-stick model of DNA and other molecules as well as the planetary model of the atom. However, as will be explained in the coming pages, these models are already outmoded.

Materialized Psychology and Biology

A materialized system of thought allows no scope for soul, spirit, and life. These entities cannot be expressed in terms of small spheres of matter. As such, it is ironic that this solid-particle mode of thought dominates the branches of science whose names mean "soul," "spirit," and "life." The word *psychology* arises from the Greek words *psukhē* and *logia*. The prefix *psycho-* means "soul" or "spirit" and the suffix *-logy* means "knowledge" or "study of." That is, *psychology* means "the study of soul or spirit." However, since physicists in the nineteenth century were unable to view or measure soul or spirit instrumentally, materialism completely denies their existence. Thus, a dilemma confronting psychology can be expressed as: "What is psychology's reason for existence if there is no soul and spirit in the world?"

The situation for biology is quite similar. The Greek root of the prefix *bio-* means "life," so *biology* means "knowledge of life." Given this, what is the attitude of biologists toward life? Biology enjoyed rapid development during the twentieth century, progressing from the study of anatomy to cellular-level histology (the study of the anatomy of cells and tissues at the microscopic level) and cytology (the study of cells' structure, function, and chemistry) and finally to the limit of the reductionist approach, molecular biology. Having achieved all these advances, biologists were quite disappointed to find no miniscule particle in the body that could be labeled "life."

The typical attitude of modern biology toward life could be represented by the comments of Albert Szent-Györgyi (1893–1986). As a Nobel Prize–winning biochemist, most famous for discovering vitamin C, he can be considered a significant authority on biology's notion of life. During an international conference in the 1970s he was asked, "What is life?" After a prolonged silence, where many around the room breathlessly anticipated a profound answer, Szent-Györgyi made a tight fist, struck the desk heavily, and said, "This is life." In fact, in his book *Introduction to a Submolecular Biology,* he had already unambiguously stated his opinion: "The biologist wants to understand life, but life, as such, does not exist: nobody has ever seen it. What we call 'life' is a certain quality, the sum of certain reactions of systems of matter, as the smile is the quality or reaction of the lips."[1]

Given that its name means "knowledge of life," biology confronts the same problem as psychology: "What is the meaning and necessity of biology if there is no life at all?" Perhaps this is the reason that many biologists have converted biology from science to technology—the technology of micromolecules and cells, namely bioengineering or biotechnology.

The pioneer of psychology, Sigmund Freud (1850–1934), whose concepts deeply influenced its development, was also a materialist. Since his time, research in psychology has focused primarily on trying to discern the material foundation of psychological and psychiatric phenomena in terms of the activity of the nervous system. This is partially correct, since any psychological and psychiatric phenomena must involve some relation to the nervous system.

Recently some biologists have begun to consider the existence of a morphogenetic field, a kind of organizing field for biological processes, or biofield. However, this research is still in a primary phase. Some psychiatrists have also started considering the existence of another world, one of invisible spirit.[2] While some serious scientific research has been undertaken in this area, it is still underdeveloped.

Spiritualized Physics

Having considered psychology and biology, what about physics? Most imagine physics as a lifeless realm composed of material and impassive particles, a world without scope for the existence of spirit, life, or soul. This view is not restricted to the mainstream public; most professional physicists also hold this opinion. Consider the following experience I had as a member of Science and Medicine Network (SMN), an interdisciplinary association of scientists and medical doctors created to discuss problems at the frontiers of medicine. At one of their meetings in Venlo, Netherlands, in 1995, a French physicist gave a presentation titled "More Spiritual Physics." The obvious implication of the title was that he considered physics to be too material, and to move beyond these limitations, that we should pay much more attention to the spiritual side of the discipline.

During the discussion that followed the talk, I commented that physics as it presently stood was actually not as material-focused as people commonly believed; it was already quite a spiritual discipline. For example, physicists talk about energy without anyone ever seeing it or presenting it for examination. It is an intangible entity. Given this, could Szent-Györgyi's attitude toward life be applied to energy? That is, there is no such thing as energy; nobody has ever seen it, therefore, there is no energy at all.

When you consider energy, its intangible nature is really akin to a kind of ethereal phenomenon or spirit. The only difference between energy and spirit is that energy is quantified by mathematical formulas and behaves in accordance with the law of conservation of energy. If we could one day quantify ethereal phenomena by means of mathematical formulas, it would become a worthy subject for serious scientific research.

In fact, physics has been a spiritual discipline from its very inception. Physics started with astronomy, the study of celestial bodies like the sun, moon, and other planets. It would seem that these huge material structures offered little room for the ethereal. However, there is an imperceptible entity at work in astronomy: gravitation, or gravity. Gravitation largely governs the movement of celestial bodies and the dynamics of the universe. Nevertheless, nobody has ever seen gravity.

Gravitation's intangible nature poses a potential challenge to some physicists in accepting its existence. William Thomson (1824–1907), a British mathematician and physicist, expressed the attitude typical of some physicists toward gravitation. He said that Isaac Newton (1642–1727) did not discover gravitation at all. Instead, Newton only discerned a similarity between the movement of an apple and the movement of celestial bodies. In other words, perhaps gravitation does not exist at all. It may merely be an abstract construct that Newton inferred from the movement of apples and celestial bodies.

The scientists in the world of the blind people could replicate the preceding attitude. Some might infer the existence of sunlight by way of comparative observations of the changes of temperature inside and outside a room. However, others would state that there was no such thing as light, since nobody had ever touched it, nobody had ever heard it, nobody had ever tasted it, and nobody had ever smelled it. Therefore, without additional supporting evidence, light could be considered a construct that some scientists inferred from the observation of temperature changes.

The key point is that the only distinction between the concepts of gravitation, energy, and those of life or spirit is that physicists have already found the quantitatively mathematical formula for gravitation and the law of conservation for energy. As yet, they have not discovered these for life or spirit. This is the reason we believe that gravitation and energy are scientific entities while life and spirit are not.

If we adopt an open mind to the existence of life and spirit, the argument can be made that these phenomena warrant future quantitative investigation. The objective of this investigation would be to discern mathematical formulas and laws of conservation for life and spirit. Achieving this would enable us to understand, manage, and cooperate with life and spirit in a much better way, namely, more rationally, quantitatively, and scientifically.

Discovering laws of conservation for life, spirit, and consciousness is, of course, challenging. Inherent in this task is the requirement to express them mathematically and quantitatively, like the laws of gravitation and energy. Perhaps it is not a job for the scientists of our generation, but I am confident it will be the job of coming generations. A brief account of the history of some advances in physics, including the discovery of the ethereal electromagnetic field, will help illustrate why we should be optimistic about this prospect.

The Study of Invisible and Untouchable Fields in Physics

People in ancient Greece believed that magnets possessed life because they attract one another. Along similar lines, the Chinese word for magnet translates as "a stone with love." Obviously, both life and love are quite spirit-like entities. However, after the long endeavor of many physicists, including decisive progress by Michael Faraday (1791–1867) and James Clerk Maxwell (1831–1879), this phenomenon was brought within reach of rational comprehension. The ghostlike entity of life and love around a magnet is now called the electromagnetic field.

Faraday, an experimental physicist, carefully observed interactions involving magnets and their relationship with electricity. In the process, he invented a simple way to visualize the invisible entity around magnets. He achieved this by putting some fine iron powder on a piece of paper placed on top of a magnet, thus enabling easier observations of the invisible entity.

Building on the experimental results of Faraday and others, Maxwell, a theoretical physicist, constructed a theoretical model that achieved a quantitative and elegant description of the movement of the imperceptible entity. Maxwell speculated that the electromagnetic field behaves similarly to liquid. He borrowed some mathematical formulas from hydrokinetics (physics concerned with fluids in motion) to describe the movement of the invisible electromagnetic field.

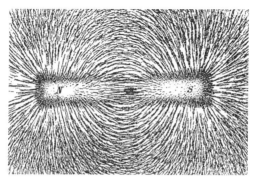

Figure 2.1. Iron powder on paper showing the field lines. Image by Zephyris and used under Creative Commons license (http://creativecommons.org/licenses/by-sa/3.0).

Maxwell's insights led him to make a remarkable prediction: that there should exist, within this intangible electromagnetic field, a kind of wave. The existence of this invisible wave was subsequently confirmed practically by Guglielmo Marconi (1874–1937), an Italian engineer. In accordance with Maxwell's astonishing prediction, he invented the first transmitting antenna and demonstrated the first wireless communication in 1901. This achievement won Marconi the Nobel Prize in 1909. Today, it is impossible to imagine modern society without the understanding and application of electromagnetic waves, which underlie mobile phones, television, radio, and satellite communication and navigation.

The discovery of yet another ethereal phenomenon, gravitation, has a similar story. Modern astronomy started with the work of Nicolaus Copernicus (1473–1543), who determined that the earth is always moving in orbit around the sun. Subsequently, the earth's orbit and those of other planets were precisely calculated by Johannes Kepler (1571–1630). This monumental achievement was made possible by the mountains of observational data gathered by his mentor, Tycho Brahe (1546–1601). While Kepler had precisely calculated the orbits, no one was able to explain the reason for them until Newton's breakthrough. The great scientist formulated an astonishing hypothesis: among the sun, earth, and other planets exists an invisible and seemingly mystical entity, which he called gravitation.

In our era, gravitation has been precisely measured and calculated. However, the existence of the gravitational wave, predicted by theories quantifying gravity, has not yet been completely confirmed experimentally, let alone applied. Nevertheless, the existence of gravitation has been widely and firmly accepted by the scientific community as one of the four fundamental forces in nature[3] alongside the ethereal electromagnetic force—the source of electromagnetic fields and waves.

Exploring the Invisible World of the Atom in Physics

While everyone today concedes the objective existence of atoms, nobody has ever seen one. Objects of this size are invisible to us—in other words, we are all "blind" in the nanoworld of the atom.

Werner Heisenberg (1901–1976) was one of the founders of quantum physics, which deals with behavior at atomic-length scales and smaller. He stated that "what we observe is not nature itself, but nature exposed to our mode of questioning."[4] This is especially applicable to the discovery of the atom, as the notion of its existence started from pure speculation thousands of years ago. It became established as a firm belief in the scientific community during the eighteenth century.

The speculation is believed to have originated from a conversation beside the Mediterranean Sea between the Greek philosophers Democritus (circa 460–370 BCE) and Leucippus about whether it is possible to endlessly divide an apple, a pear, or a strand of hair into smaller and smaller pieces. Their conclusion was that the process could not be endless. At some definite moment, they believed they would encounter the smallest particle that still possessed the properties of an apple, pear, or hair. They also speculated that everything in the world is made of these smallest possible invisible particles, which they called atoms. The word *atom* stems from the Greek word *átomos,* which means "indivisible." Between these particles there is a void—a complete absence of any matter.

At the time, this idea could not proceed beyond pure speculation; no one could devise a practical experiment to judge whether it was correct or not. It was largely forgotten until the seventeenth century, when the French philosopher and astronomer Pierre Gassendi (1592–1655) recalled the hypothesis. Newton subscribed to the theory, even though there was still no experimental evidence. Following its reintroduction, the existence of the indivisible particle became a core belief in the scientific community, and many scientists tirelessly sought evidence for it.

Daniel Bernoulli (1700–1782), a Dutch mathematician, is credited with the first significant step toward scientific and quantitative recognition of the atom.

Figure 2.2. The ball-and-stick models of a water molecule and a sugar molecule. See plate 2.

He proposed that gas consists of many tiny invisible balls, moving at high speeds, that constantly collide with one another. All phenomena concerning the temperature, pressure, and expansion of gases could easily be explained in terms of these collisions. In 1783, the English scientist Henry Cavendish (1731–1810) and French chemist Antoine-Laurent de Lavoisier (1743–1794) discovered that air was not a pure substance but instead a mixture consisting mainly of nitrogen and oxygen. In addition, they also proved that carefully burning a mixture of two parts hydrogen and one part oxygen would produce pure water without any remaining gas.

The next significant advance came in 1803, when English chemist and physicist John Dalton (1766–1844) formulated the law of material composition. It stated that a chemical compound consists of pure materials in definite proportion. In essence, this established a chemical theory of the atom and formed the foundation of modern chemistry. In light of the law of material composition, Dalton postulated that the atom looked like a ball, and that a molecule looked like a combination of many balls. A water molecule would look like a combination of three balls: a big one, the oxygen atom, and two small ones, the hydrogen atoms. In blind people's terms, this was the first model, or hypothesis, or theory concerning the appearance of the "tiny, invisible, and untouchable elephant."

These balls, or particles, are inconceivably tiny. In the nineteenth century, however, the Italian physicist Amedeo Avogadro (1776–1856) and the French physicist and mathematician André-Marie Ampère (1775–1836) precisely calculated that there are 602,000 billion billion atoms in a gram of water and 25 billion billion atoms in a gram of air. This achievement granted the product of Democritus and Leucippus's speculation a quantified basis of proof within the viewpoint of modern science.

Invisible Rainbow

In addition to its importance in developing our understanding of the atom, Dalton's work firmly established the basis from which modern chemistry developed. As such, he is credited with being its founder. The ball-and-stick model of the molecule (fig. 2) that he established is able to describe the structure of any molecule successfully and clearly, and currently it dominates not only chemistry but also biology and physiology. It also forms the basis of the modern pharmaceutical industry and Western medicine. Due to its immense success, the ball-and-stick model has become such an entrenched notion that many people believe it to represent reality. However, as the coming pages will explain, this is not the case.

The great success of chemistry in the nineteenth century confirmed the particle pattern of the world. Subsequently, the development of quantum physics in the first half of the twentieth century, in particular the explosion of the nuclear bomb in 1945, cemented the absolute authority of this view. The discovery of the DNA double helix in 1950 brought molecular biology to the forefront. The DNA molecule was shown to encode all the genetic information for the development

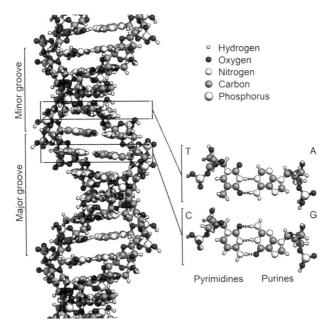

Figure 2.3. A section of a DNA molecule. Image by Zephyris and used under Creative Commons license (http://creativecommons.org/licenses/by-sa/3.0).

and functioning of known living organisms. Thus, even the secret of life could be found in sequences of chemically bonded atoms.

Encouraged by molecular biology's success, psychologists began to search this particle perspective of the world for scientific explanations for emotional, and what had previously been considered spiritual, phenomena. For instance, anger could somewhat be attributed to the presence of excess adrenaline and happiness partially ascribed to the presence of morphine. Recent speculation has even considered the existence of a "consciousness particle," which would provide a scientific explanation for consciousness in terms of particles.

The continued preponderance of the ball-and-stick model is evidenced by the fact that, already being dominant in biology, it has become a prevailing influence in psychology. Many psychologists today seek the soul and spirit within its constructs. This pattern of development explains the first part of Stapp's statement that opened this chapter: "While psychology has moved toward the concepts of nineteenth-century physics, physics itself has moved in just the opposite direction." As the second part of the statement implies, many modern physicists now understand that the speculation of Democritus is not the true picture of our world.

The Atom Is Not the Fundamental Indivisible Particle

Toward the end of the nineteenth century, scientific developments, including the discovery of the electron, had already revealed that the atom was not the fundamental indivisible particle; it possessed its own detailed internal structure. In other words, closer inspection of the ball-and-stick model reveals that the atom cannot be considered a hard, solid sphere, instead composed of a relatively complicated substructure of component particles that are perhaps themselves composed of even smaller fundamental particles.

The recognition of the substructure of the atom developed in several stages. Joseph John Thomson (1861–1933), a British physicist, proposed the first model of atomic structure in 1898. He theorized that the atom resembled a round ball in which the positive material was continuously distributed with negative electrons scattered separately within the positive matrix. A British dessert called plum pudding served as an analogy to describe the model, with the electrons likened to plums scattered within a positively charged pudding. Consequently, this model was referred to as the plum-pudding model.

In 1904 Thomson's model of the atom was found to be inconsistent with new experimental data. This led Ernest Rutherford (1871–1937), another British physicist, to propose the planetary model of atomic structure: a miniscule nucleus in the center of an atom contained most of the positive electrical charge and mass. The electrons, with their negative electrical charge, rotated around the nucleus like a tiny solar system.

Danish physicist Niels Bohr (1885–1962) improved Rutherford's model of the atom in 1913. He proposed that electrons were restricted to some special stable orbits that they could occupy without loss of energy (the third image in plate 3 in color plate section). His model of the atom, the Bohr model, is what the majority of scientists and other people visualize when they consider an atom.

Despite its mainstream popularity, Bohr's model does not represent the final refinement. German physicist Max Born (1882–1970) postulated that the structure of the atom could be visualized as an electron cloud surrounding the nucleus (the fourth image in plate 3 in color plate section). This is now the accepted visualization of an atom for physicists.

These developments can be summarized as follows. The atom envisioned by Democritus and Leucippus and firmly proven by Bernoulli, Cavendish, de Lavoisier, Dalton, and many other scientists was in fact not the fundamental and indivisible particle. It was, at a minimum, composed of electrons, protons, and neutrons. Subsequently, many physicists labeled these as basic particles, believing them to be fundamental and indivisible.

No Fundamental and Indivisible Particle: Nothing Is Real but Vacuum

Unfortunately for Democritus and Leucippus, further developments in physics came to show that their speculation was completely inaccurate from the very beginning. That being said, the quest to find the fundamental particle of the world provided a strong motivation to develop science and led to fruitful results.

According to their speculation, the world is made of indivisible particles separated by a void. However, even in the early stages of the development of quantum physics, some uncertainty about the particle pattern of the world existed. The first prominent rebel against the dominating hypothesis of Democritus was renowned physicist Albert Einstein (1879–1955). He predicted that physical matter could be transformed into intangible energy. His famous formula, $E = mc^2$, essentially equates energy to matter. In fact, the working principle behind the nuclear bomb is based solely on this transformational relationship between matter and energy.

British physicist Paul Dirac (1902–1984) instigated the second notable rebellion against the dominant particle hypothesis of Democritus. Dirac did this by predicting the existence of the antielectron—a particle with the same mass as an electron but with the opposite electrical charge. His prediction stated that, when an electron met an antielectron, both of them would disappear into a vacuum. This prediction was soon experimentally confirmed, and Dirac was awarded a Nobel Prize in 1933.

Soon afterward, discoveries by many other physicists led to the realization that all particles have corresponding antiparticles. In other words, all the matter that makes up our world has corresponding antimatter. If all these particles were to meet with their respective antiparticles, the whole world would disappear into a vacuum—a complete absence of matter—through the annihilation reaction between particle and antiparticle (particle + antiparticle = vacuum).

The reverse process of the annihilation reaction is also valid and is called pair production, in which a particle and antiparticle pair can be created from a vacuum by means of stimulation by radiation. The process involves high-energy gamma rays converting into the particle-antiparticle pairs. In other words, the vacuum is actually also an ocean full of matched particles and antiparticles. For this reason, some physicists call the vacuum the "Dirac Ocean."

Nowadays, more and more physicists accept the viewpoint that all matter that constitutes the material world is, in fact, only a vacuum fluctuation, nothing more. The particles that we regard as tiny solid spheres are in reality only hard cores of energy, or tight wave packets. It is possible to envision the nature of a hard core of energy by using the analogy of an electric fan. When the fan is rotating at high speed, you cannot throw a ball through the rotating blades. It becomes a hard plate, and the tactile sensation of making contact with it would be akin to a solid plate rather than a piece of platelike energy.

The implications of this understanding of reality become somewhat unsettling to imagine when we analyze any object to its reductionist end point. Consider the example of an ordinary desk. In line with our knowledge of molecules and atoms, we know that the wood is made of molecules, mainly cellulose. Cellulose is composed of glucose, which is in turn composed of atoms of carbon, oxygen, and hydrogen. These atoms actually occupy only a very small portion of the space in the wood of the desk—the rest is empty. If an atom could be magnified to the size of a football field, the nucleus, which contains virtually all the mass, would be only as big as a football. The electrons surrounding it would be even smaller. Consequently, with the exception of the tiny proportion of space occupied by their subatomic particles (electrons, protons, and neutrons), atoms are entirely empty. What appears to be a desk made up of matter is in fact mainly empty space. Further examination of the subatomic particles would reveal them to be only hard cores of energy.

When the process of reductionism is pursued to its end point, there exists only immaterial energy, or vacuum—nothing else. It is provocative that as the development of modern physics has progressed deeper and deeper, it has eventually approached the spiritual dimensions of the ancient, most fundamental motto of Buddhism: "Material is emptiness and emptiness is material." Similar ideas can also be found in the Bible: "What can be seen was made out of what cannot be seen,"[5] and, "What can be seen lasts only for a time, but what cannot be seen lasts forever."[6]

To summarize, the hypothesis of Democritus concerning the indivisible atom does not describe reality, and particles do not form the essence of the world. Modern physics rejected this hypothesis more than one hundred years after science initially confirmed it.

Unfortunately, the dissolution of Democritus's dream is an extremely unsettling process. Its impact has been felt not only by ordinary people but also by most scientists, particularly those invested in the approach of reductionism. Einstein's discovery of the law of transformation between matter and energy greatly undermined the basis of materialism. The loyal Marxist and materialist Vladimir Lenin (1870–1924) was among the first people to become aware of the danger of this law of transformation. In order to rescue the basis of materialism, in 1908 he wrote a book titled *Materialism and Empirio-criticism,* in which he created a new definition for the term *material:* "The sole 'property' of matter with whose recognition philosophical materialism is bound up is the property of being an objective reality, of existing outside the mind."[7] Using such an extensive definition facilitated the inclusion of energy into the area of matter. However, he was not aware that within his new definition, consciousness is beyond, over, or even above matter. Therefore, he was not successful in rescuing materialism.

It is interesting that most physicists were much less sensitive than the politician Lenin to the implications of the major changes in the concepts of physics until the appearance of Fritjof Capra's popular book *The Tao of Physics* in 1975. Naturally, biologists and psychologists are even less sensitive than physicists to developments in the concepts of physics, and most of them, particularly molecular biologists and molecular psychologists, operate completely within the concepts of nineteenth-century physics. Meanwhile, physics itself has moved in exactly the opposite direction, as Stapp pointed out in the statement that introduces this chapter.

From a Hard Core of Energy to Dispersed Energy

During the first two decades of the twentieth century, modern physics proved that Democritus and Leucippus's view of the world was incorrect. Physics invalidated the hypothesis that the world is composed of the smallest invisible particles and that between these particles is emptiness. The concepts of modern physics hold that everything in the world is only vacuum fluctuation. In other words, the field and the wave are the essence of the world. The particle that Democritus and Leucippus speculated about is not the essence of the world, but instead a sort of wave package or hard core of energy. Between these hard cores of energy is dispersed energy. This is the medium that links the hard cores of energy to form the huge network of the universe.

Notwithstanding its ultimate invalidation, we must acknowledge that the particle pattern provided the first stage to recognize the world, and it was extremely useful. Since then the study of science has progressed, from visible to invisible, from tangible to intangible, and from the hard core of energy, namely the particle, to the diffuse energy pattern, namely the electromagnetic field. In other words, Democritus represents the primary stage of science, as it is relatively simpler to study the hard core of energy than it is to study the dispersed energy between these hard cores.

CHAPTER 3

INAUDIBLE MUSIC AND THE INVISIBLE RAINBOW AROUND US

> We automatically assume that our senses can give us a complete picture of our surroundings, and it seems to us strange, at first, to be considering the causes of influences and phenomena that we cannot perceive in this way.
>
> —HERBERT L. KÖNIG

Now it is time to move beyond the invalidated viewpoint of Democritus. Advancing from the indivisible particles of matter that underpinned his worldview, we will progress to a more fundamental outlook. In doing so we will encounter a more fundamental aspect of biology, psychology, and medicine.

As König notes above, we automatically assume that our senses can provide a complete picture of our surroundings. This assumption, even belief, is evidently untrue. Unfortunately, most people, including many well-educated scientists, overlook the fallacy of this assumption, instead firmly holding to the belief that our senses already give us a complete picture of our world. For example, many biologists, even psychologists, firmly believe that our senses, aided by the technology of the electron microscope and chromatography, a range of techniques that separate mixtures into their constituents, already give us a complete picture of living systems, such as our bodies and cells.

The present situation of biology can be summarized as follows. Molecular biology's rapid development in the last half century has led biologists to firmly believe that every detail of the cell is well known. In some ways, modern biologists are as familiar with the cell as you are with your own living room, or even more so. You know the structure of your living room and all the furnishings. Suppose that your familiarity even extended to the level of detail en-

Inaudible Music and the Invisible Rainbow around Us

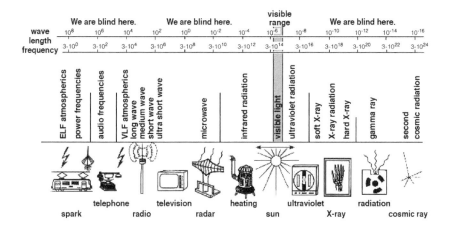

Figure 3.1. The spectrum of electromagnetic waves.

joyed by a modern biologist. Then you would be aware of every molecule in your living room, including those in the furnishings, in the walls, and in the air. At this point, it would appear that you possessed a complete picture of the living room.

Then again, if you turn on your radio, music comes out. Evidently, the music must have been there already, like a ghost singing inside your room. If you turn on your television, a vivid, dynamic scene is portrayed. Again, the scene is unfolding somewhere in your room, like a cohort of immaterial beings playing and dancing. As with radio, the ethereal assembly already occupied your room before you switched on the television.

Conventional opinion in the modern world no longer allows for the existence of ghosts. They are considered a superstition that only existed as a mainstream belief during uninformed times in the distant past. Modern people proudly put their faith in science alone: faraway radio stations produce music, and television stations are the source of the dynamic scene. While this is true, the signals already surround you as you sit in your room. A simple indoor antenna facilitates their easy reception. Thus, even if you charted every last molecule in the room, you would not have a complete picture of your living room. To put it another way, this awareness of every single particle would reveal, at most, half the picture. The other half is invisible, very ethereal, and beyond the limitations of our senses.

Modern science has revealed that visible light is a type of electromagnetic wave. The electromagnetic spectrum, as shown in figure 3.1, organizes these waves from left to right in order of increasing wavelength and decreasing frequency. A brief inspection of this spectrum reveals the feebleness of our ability to sense these waves. Visible light represents a very small part of the spectrum.

The frequencies of electromagnetic waves in this spectrum range from 0.5 hertz to 3×10^{24} hertz. A hertz (Hz) means a complete cycle, or wave, per second. Thus the spectrum ranges from half a wave per second to 3,000,000,000,000,000,000,000,000 waves per second. In terms of wavelength, the spectrum ranges from 1,500 meters to 10^{-16} meters (0.0000000000000001 meters). Within this, the very narrow section of visible electromagnetic waves range only between 360×10^{-9} meters and 760×10^{-9} meters (the gray range in fig. 3.1).

Consequently, the invisible part of your living room is, in fact, much larger and richer than the visible and molecular part of the room. This is because the information carried by waves is much more abundant in the invisible part than in the visible part. When it comes to this greater part of the world, we are all, unfortunately, blind. Our situation is that of the blind people discussed in chapter 1. Their immense challenge was to achieve awareness and come to accept that their world was so colorful.

Modern people, with their faith in science, refer to this invisible part of our world as the invisible range of electromagnetic waves. Perhaps people in ancient times, living before the development of modern science, would have labeled it the ghost world. Of course, the invisible range of electromagnetic waves and the ghost world are not identical, but there is at least some overlap between the two concepts. To illustrate, let us imagine what we would observe if we possessed the supernatural ability to see the invisible range of electromagnetic waves.

The Invisible Rainbow, or the Ghost World

Acquiring the ability to view the invisible range of electromagnetic waves would be a disturbing experience. Everything in our surroundings would appear different. The familiar everyday world would be transformed into an alarming phantom world.

To start with, the colors of almost everything would be different. As is evident in figure 3.1, the electromagnetic waves emitted from the sun are not limited to the visible range. Therefore, the color of the sun would be a strange, previously unseen "color," because the observer would interpret these previously undetected electromagnetic waves as new colors, and they would remain visible even behind thick clouds.

Inaudible Music and the Invisible Rainbow around Us

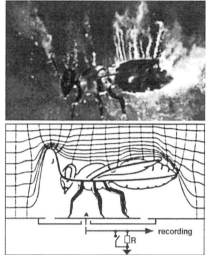

Figure 3.2. Thermograph around a human body (left) and the electrical field around an insect (right). See plate 4.

Many fish, reptiles, and insects are able to see ultraviolet light. Bees, for example, use light in the ultraviolet range to distinguish and locate flowers. Animals continuously emit infrared radiation; this is how a snake can detect a mouse in the dark of night. Thus all the colors of the animals would also have changed. Perhaps the most unsettling part of the experience would be the dramatic changes to the faces of those around you. In addition to the visible light normally present, their faces would also be reflecting ultraviolet rays, infrared rays, microwaves, and so on.

Scary faces aside, you would also find that the people you know would have taken on a somewhat holy appearance. Everyone you looked at would be encircled by an aura (fig. 3.2 above and plate 4 in color plate section), much like the one that ancient cultures believed only holy people possessed. In truth, the appearance of an aura does not imply holiness. Any living creature has an aura surrounding it. Auras are not even limited to living creatures: many nonliving objects, such as heating radiators, are encircled by a strong infrared aura, similar to the image on the left in figure 3.2. Even houses with temperatures only a little higher than their surroundings have an infrared aura.

A radio station's antenna is encircled by longer electromagnetic waves. If the range of our supernatural vision extended to radio waves, we would see an aura

31

Figure 3.3. The aurora borealis, or northern lights, shine above Bear Lake, Alaska.

around the antenna dancing to the rhythm of the music. If we had the ability to hear the rhythm of electromagnetic waves, as we can hear acoustic rhythm, we would hear the music directly from the antennae of analog radio stations. Similarly, we would also see a pulsating aura around every mobile telephone and, in the case of analog communication devices, hear the conversation without the need for a handset. In the case of digital devices, while we could hear the transmission, it would be in the form of an unintelligible digital signal.

A car would no longer appear in the shape and colors that we are used to seeing. Instead, it would be surrounded by a very complicated aura, formed from a range of infrared rays, microwaves, shortwave radio, longwave radio, audio frequency, power frequency, and extremely low-frequency (ELF) waves.

The backdrop of our regular reality, the blue sky, would also appear very different. We would see cascades of luminosity descending from outer space, composed of waves generated by particles being created and destroyed, in rhythmic pulses. The ultimate source of the energy behind this phenomenon is the solar wind—a flow of plasma emitted from the sun in all directions. Plasma is a state of matter that no longer exists as atoms but as charged particles, and the electrons move around separately from the nuclei. Most of the charged particles that constitute the plasma are deflected by the earth's magnetism, with the exception of the two polar areas. Here the creation and the destruction of particles is so intense that it generates electromagnetic waves in the visible range, perceptible to the naked eye; we know this phenomenon as the auroras.

Once gifted with this special ability, we would immediately encounter a major challenge in communicating our experience. We would find that the language required to convey our experience of seeing the invisible range of electromagnetic waves, the "ghost world" of ancient times, does not exist.

As discussed, we are only able to see a small range of electromagnetic waves with the unaided eye, from 360×10^{-9} meters to 760×10^{-9} meters. In other words, this is the only region where we are superior to the blind people of chapter 1. Within this narrow visible range, our ancestors invented words, called colors, for special divisions of the visible range:

360–430 nm	violet
430–455 nm	indigo
455–492 nm	blue
492–550 nm	green
550–588 nm	yellow
588–647 nm	orange
647–760 nm	red

Aside from these seven words naming the basic colors visible in a rainbow, our ancestors also invented many other words for mixes of different proportions and intensities of these colors. This multitude of terms facilitates the description of our colorful world in reports, novels, and poetry.

Our ancestors did not invent appropriate words to describe the electromagnetic world. Modern scientists, particularly physicists, have had to invent new words, including *ultraviolet ray, infrared ray, microwave, shortwave, longwave, audio frequency, power frequency,* and *extremely low frequency,* to describe the colors in the invisible region of the electromagnetic spectrum. Regrettably, these special terms are not capable of conveying any sense of beauty and instead often induce headaches. This makes the challenge of conveying the charm of the invisible world akin to that confronting the scientists in the world of the blind people:

how to describe the beautiful rainbow to others without the words *red, orange, yellow, green, blue, indigo, violet,* or any other related to color or light.

Any scientist working on the frontiers of research will encounter the challenge of language. Heisenberg writes, "It is not surprising that our language should be incapable of describing the processes occurring within the atoms ... it was invented to describe the experiences of daily life."[1] While it is understandable that terms along the lines of *audio frequency* and *dissipative structure* might evoke loathing, imagine that they are elegant words describing beautiful colors. This will enable you to begin to imagine, and even appreciate, the beauty of this ethereal world.

Wave Interference

The realm of waves exhibits some odd phenomena that do not exist in the domain of molecules. For example, while it is impossible for two objects or two molecules to occupy the same position at the same time, it is quite normal for two, or even many, waves to do so. In some ways, this is analogous to the following comparison: while two people cannot physically occupy the same position at the same time, ancient belief allows for the possibility of two ghosts simultaneously occupying the same position. The difference between ancient people and modern scientists is that people in ancient times achieved perception beyond their senses through their intuition and engaged poetic language to describe their experiences, like describing a dream. In contrast, modern scientists discern the ghostlike invisible waves with instruments, mathematical inference, and calculations. They then describe it with rational, rigorous, and quantitative language. Ancient people did not specify what would happen if two ghosts sat in the same chair at the same time. Modern scientists can clearly state what happens when two waves occupy a place at the same time, and can even precisely calculate the result of the co-occupation of two, or even many, waves.

As figure 3.4 illustrates, at places where the peaks of two waves meet, they add together and form an even higher wave. Conversely, at the place where the peak of one wave meets the trough of another, they cancel each other out. Therefore, the result of two waves being superimposed is definite and can be calculated precisely. Of course, scientists won't refer to this as the result of two ghosts sitting together. Instead, it is described as the "superposition of two waves," or the "interference of two waves." In addition to being more precise, the use of such scientific terminology also invokes an air of authority.

Inaudible Music and the Invisible Rainbow around Us

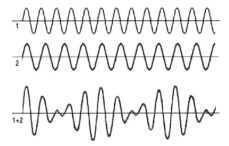

Figure 3.4. Interference of two waves.

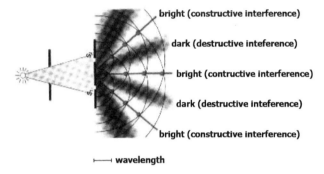

Figure 3.5. Interference of two light beams.

Interference is an important phenomenon in physics and can be either constructive or destructive. Figure 3.5 depicts a situation where a single light beam has been divided into two by passing it through two narrow slits, S_1 and S_2, in a screen. The two resultant beams emanating from these slits have then interacted with each other. In regions of destructive interference, the two beams cancel each other, resulting in a darker area. In regions of constructive interference, the two beams strengthen one another and form an even brighter area.

The experiment shown in figure 3.5, called the double-slit experiment, can be performed in our everyday visible range of the electromagnetic spectrum. However, the phenomenon of interference also occurs in the ethereal invisible range of electromagnetic waves, and makes this otherworld even more abundant and "colorful."

Aside from the scenario of the double-slit experiment, typical of a physics student's experience, a more common example of interference can occur on puddles in the street after a rain. If a layer of oil happens to coat the puddle's surface when the sun reemerges, a colorful pattern will appear, attributable to the phenomenon of interference within the visible range of electromagnetic waves, or visible light. If the expansive invisible range of electromagnetic waves were sampled, the phenomenon would be much richer.

The Invisible Structure of an Interference Pattern

It is important to highlight a further implication of the phenomenon of interference. In addition to bringing more vibrancy to the realm of waves, it also generates new structures. On initial consideration, it sounds improbable that such dynamic entities can create a stable structure. Nevertheless, any fast-moving wave can form a stable "standing wave" when it traverses back and forth many times between two boundaries.

Figure 3.6 displays an example of a mechanical wave on a string. The two ends limit and reflect the wave back on itself. The backward wave is added to, or superimposed on, the forward wave, resulting in destructive interference in some areas and constructive interference in others. As a result, some points on the string never move, and at other points the maximum and minimum fluctuations always occur. This results in a stationary pattern, as shown in figure 3.6, even though the wave is in constant rapid motion.

In mathematical terminology, the example above is a type of one-dimensional standing wave. The same principles also allow two-dimensional and three-dimensional standing waves. Figure 3.7 shows a variety of two-dimensional standing waves that can be established at different frequencies in a disk. In comparison to the one-dimensional variety, the configurations of two-dimensional standing waves are usually much more complicated. Despite being composed of constantly moving waves, standing waves are not only patterns but also relatively stable structures.

The two preceding examples illustrate visible stable interference patterns. Many unseen stable interference patterns also occur. For instance, the elegant

Figure 3.6. One-dimensional standing waves on a string.

Inaudible Music and the Invisible Rainbow around Us

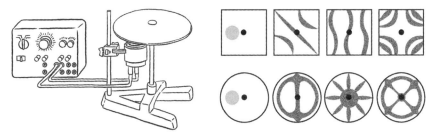

Figure 3.7. Two-dimensional standing waves and stable interference patterns.

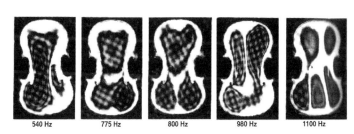

Figure 3.8. Two-dimensional standing waves and stable interference patterns in a violin.

contours of a violin serve not only for appearance but also for enhanced acoustic effect. The "sound color" of the instrument is actually the amalgamation of many special frequencies combining in overtone. The shape of the resonance cavity selects these frequencies. In a given cavity shape, waves of certain frequencies can be reflected in such a way that they form stable patterns of interference. Waves of these frequencies will absorb energy from other frequencies and last much longer. Figure 3.8 shows how a number of frequencies create stable interference patterns of standing waves in a violin. In the three-dimensional case, the interference pattern of standing waves forms even more complicated stereoscopic structures that remain dynamic at the same time.

Inaudible Music: Ghost Music

It is worth noting that while the acoustic standing waves in a violin are invisible, they are audible, allowing us to enjoy the music from the violin and appreciate its beautiful "sound color." However, our listening ability is limited, sampling the relatively narrow range of 20 to 20,000 hertz. Most mechanical waves exist at frequencies above this threshold. As such, they are referred to as ultrasonic waves and are both invisible and inaudible to us.

In addition to being unable to appreciate the virtual symphony of mechanical waves beyond the audible range, we are also completely deaf to music in the form of electromagnetic waves. Despite being rationally aware that the electromagnetic waves from radio stations surround us with music, we are unable to experience it directly. We can experience it with the help of a radio, but only one channel at a time. This represents a very small part of the music unfolding in the electromagnetic realm. The humbling truth is that we are both blind and deaf to most of the world.

Fortunately, it is not impossible for us to gain a sense of, or even appreciate, the music in the electromagnetic realm. At some concerts, colorful lights in the theater are modulated by the rhythm and melody of music. This is one way to make invisible music visible. Similarly, telephone receivers and speakers make inaudible voices and music audible. The images in figure 3.8 were created by scientists using an interference meter that generates a visible representation of the invisible structures of acoustic standing waves.

Medical doctors already employ instruments to perceive invisible electromagnetic waves in the body in order to gather information from the heart and the brain. These techniques include electrocardiography (ECG) and electroencephalography (EEG). ECG is already over one hundred years old and is used today as a routine method of diagnosis. But few doctors would be aware that its use essentially involves reading a transcript of inaudible music with their eyes instead of listening to and appreciating it with their ears.

The Ghost Structure and Medicine

These days a fresh vocabulary has appeared and is becoming increasingly popular in books concerning energy medicine, vibrational medicine, and information medicine. Energy medicine is a branch of complementary medicine that holds that the healer can transmit healing energy into the patient. Vibrational medicine claims to transfer various forms and frequencies of vibrations into the human body and its energy field to heal the patient. Information medicine asserts that illness arises from miscommunication between the cells and seeks to rectify this.

Energy, vibration, and information can also be considered ethereal entities. They are not as easily understood and controlled as the molecular aspect of the human body that biologists presently focus on. Consequently, many scientists, particularly molecular biologists and neurophysiologists, are unable to find a scientific explanation within the materialist framework of biology and are thus opposed to alternative forms of medicine.

Invisible structures of electromagnetic waves exist not only in the inanimate world but also inside living systems, including human bodies. In the body, the vibrations of organs, tissues, cells, and molecules generate the electromagnetic waves that interact to form many invisible structures. In turn, they determine the distribution of energy inside and around a body and transfer information at very high speed. As a result, this intangible world within the human body is related to the many mysterious phenomena found in complementary medicine. However, it is difficult to discern their scientific basis from the current textbooks. The next part of this book will concentrate on the relationship between these intangible structures and complementary medicine.

PART 2

**TWO PARADIGM SHIFTS—
ONE IN MEDICINE, ANOTHER IN SCIENCE**

CHAPTER 4

MAJOR CHANGES IN THE MEDICAL MARKET

A story is told in China about a peasant who had worked as a maintenance man in a newly established Western mission hospital. When he retired to his remote home village, he took with him some hypodermic needles and lots of antibiotics. He put up a shack, and whenever someone came to him with a fever, he injected the patient with the wonder drugs. A remarkable percentage of these people got well, despite the fact that this practitioner of Western medicine knew next to nothing about what he was doing. —TED KAPTCHUK, *The Web That Has No Weaver*

The current era has witnessed a revival of many archaic forms of medicine, including acupuncture, ayurveda, and homeopathy. To the surprise of many, particularly medical doctors, this renaissance of traditional therapies has begun to challenge the preeminence of Western medicine. Similarly, some scientists are privy to an uprising against the dominant frameworks of biology and physics. The parallel rebellions reinforce each another: the first is external in nature while the latter is introspective, deep, and fundamental.

The Meaning of Alternative

The outward rebellion involves so-called "alternative" or "complementary" medicines. The label "alternative" can evoke connotations of opposition that is outside government, bereft of power, apart from the mainstream, and unorthodox.

However, these medicines were not always considered alternative and were orthodox, traditional, and mainstream in Europe, the Americas, China, and India for thousands of years. The twentieth century saw them displaced by the dominance of biomedicine, marginalized and banished from the realm of mainstream medical practice. The decline of these systems of medicine can be attributed to a number of factors.

The first was their inability to decisively control bacterial and viral diseases. This was particularly true of infectious diseases, including smallpox, bubonic plague, cholera, and typhus. Here, modern Western medicine provided a definitive victory with the invention of various vaccines. Subsequently the invention of sulfonamide, the first antimicrobial sulfa drug and precursor to modern antibiotics, saved innumerable lives. Finally the invention of antibiotics conquered almost all the bacteria-based diseases. Today, while drug-resistant bacteria are on the rise and influenza pandemics still pose a risk, pandemics of bubonic plague, cholera, and smallpox that once invoked dread exist for the most part only in history. This profound contribution established a strong foundation and formidable reputation for modern Western medicine.

Western medicine's supremacy in surgery also contributed to the demise of the older medical systems. War has frequently afflicted humanity, and its injuries have contributed to surgery's perpetual importance. This was particularly evident in the two colossal wars of the first half of the twentieth century, where surgery saved innumerable military and civilian lives. During times of peace, surgery also saved many lives from injuries arising from accidents and various life-threatening health conditions.

A third major reason for the rise to supremacy of Western medicine was the great success of modern science. Its discoveries enabled the development of sophisticated techniques and technologies that changed the world and significantly increased the standard of living. Along with the success of its applications, science developed a self-consistent theoretical system that was capable of explaining most phenomena encountered in life. It can be argued that this system represents the most profound edifice of knowledge ever created by humanity.

The immense success of modern science made it the sole criteria by which everything, including medicine, was assessed. If a medicine was consistent with the framework underpinning the theoretical system of modern science, it was considered "scientific" and thus reliable and orthodox. Conversely, medical systems that did not concur with the framework, including homeopathy, acupuncture, and ayurveda, were regarded as "unscientific" and could, at best, be labeled "alternative." Given the supremacy of the scientific outlook, even the "alternative" label could be considered a generous concession.

Faded Glory

Western medicine has not been favored to the same degree since the mid-twentieth century. Surgery, while remaining important, lost the prominent arena that the major wars of the twentieth century afforded it. Similarly, technology's rapid development has seen injuries in vehicle accidents and factories greatly reduced, further diminishing surgery's prominence. As people live longer, they look to medicine to improve their quality of life. In this context, surgery offers limited assistance.

Another modern-day phenomenon undermining the dominance of Western medicine is the relative rarity of epidemic diseases. A combination of factors, such as improved hygiene and the effectiveness of earlier vaccination programs, has all but eliminated epidemics from most people's lives. While contagious diseases such as influenza, malaria, and AIDS still pose a risk to life, most people have not personally experienced a highly contagious, highly lethal epidemic. Modern medicine has not been able to provide decisive cures for many of the diseases that still threaten us and thus has not enjoyed the same opportunities to save the masses that it experienced in earlier times.

In the absence of widespread life-threatening acute diseases, inhabitants of developed countries continue to suffer from chronic diseases and functional disorders. These represent an Achilles heel for modern Western medicine. Consider the following widespread scenario: you visit your doctor suffering from a recurring headache or some other discomfort. The doctor performs a thorough investigation and refers you for an array of medical tests that employ impressively advanced technology. In many cases these tests reveal the cause of your discomfort, but often the results are within accepted parameters. This outcome leads your doctor to conclude that, contrary to what you may believe, there is nothing wrong with you.

Naturally, many people finding no remedy in Western medicine will look for something "alternative." For this reason, older medical systems such as homeopathy, acupuncture, and ayurveda have experienced a revival and a return to the medical market. A 1993 report in the *New England Journal of Medicine* found that over one-third of the 1,539 adults surveyed had used alternative therapies in the past five years, averaging nineteen visits per person. The article calculated that the nation's out-of-pocket expenses for alternative medical care amounted to around $10 billion per year. This was only $3 billion less than the out-of-pocket expenses for all hospitalizations in the United States.[1]

This $10 billion represented a major market opportunity for medical centers, and many that had previously espoused mainstream science's opposition to alternative medicine began to provide alternative, traditional, and even spiritually based therapies. Recent data from the United States shows that seventy-five medical schools now include courses in complementary and alternative medicine (CAM) in their curricula. This revival is further demonstrated by community hospitals and academic medical centers opening new services, such as the Division of Complementary Medicine at the University of Maryland, the Center for Integrative Medicine at Thomas Jefferson University in Philadelphia, and the Center for Alternative Medicine and Longevity at the Miami Heart Institute. The latter promotes twenty-six varieties of alternative therapies, including applied kinesiology, bio-oxidative therapies, Qigong, and iridology.[2]

Heading East, out of Curiosity

A curious modern development in medicine has seen thousands of well-educated Western doctors heading to China, India, and Japan to study Eastern medicine. While the early stages of this pilgrimage were mostly fueled by inquisitiveness, a desire to achieve better outcomes for their patients as well as gaining a competitive advantage in the Western market have served as more recent motivations.

Widespread awareness of Eastern medicine in the West is relatively recent. While historians have discovered records documenting traditional acupuncture treatment in China dating back five thousand years, references to acupuncture treatments did not appear in Western medical literature until the early 1800s. When first reported by Western explorers, these treatments were depicted as exotic, primitive, and unworthy of serious attention. Consequently, acupuncture disappeared from scientific and medical consideration.

In the 1920s a team of forty German doctors journeyed to China to embark on a serious investigation into Traditional Chinese Medicine. They studied both herbal medicine and acupuncture. In that era, war, starvation, and epidemic diseases still posed the most important problems for medicine, and Western medicine, equipped with its unfolding discoveries of sulfa drugs, vaccines, and antibiotics as well as its effective surgical techniques, offered more powerful solutions than traditional herbal medicine and acupuncture. Traditional therapies were also incompatible with contemporary science and technology, which had enjoyed great success during the industrial revolution and the colonial wars. In this light, Western doctors and scientists afforded little credence to exotic ancient medical therapies. Around this time, Traditional Chinese Medicine's "unscientific" label led the Chinese government even to consider forbidding it.

Later, after World War II, some pioneering Western doctors recommenced investigation of acupuncture. Employing modern technology, they focused on electronic measurements (explored in more detail in chapter 6). The first discovery occurred in 1947 when a German physician, Richard Croon, found a correlation between resistance on the skin and acupuncture point locations: areas of the skin where electrical resistance was measured to be lower were more likely to be located at acupuncture points. Subsequently, in 1950, Yoshio Nakatani, a Japanese doctor, independently found the same lower-resistance phenomena along the Kidney Meridian. In Traditional Chinese Medicine, a meridian is the path along which the Qi, or life force, flows. The Kidney Meridian is one of the twelve standard meridians; there are also eight extraordinary meridians.

From 1953 onward, another German physician, Reinhold Voll, devoted himself to systematically studying the relationship between acupuncture systems and lower resistance on the skin. Based on the results of his studies, he developed a system of using the resistance measurements for diagnosis, thereby transforming acupuncture from a purely therapeutic technique to a combined therapeutic and diagnostic system.

Despite the introduction of modern techniques to acupuncture, thereby making it more powerful, it still lacked scientific rationale to explain its function. Acupuncture was not accepted by the scientific community, medical insurance, or medical universities, even in Germany, where Croon and Voll worked.

Heading East, for Business

The profit motive is generally a more powerful driver than curiosity. The evolution of the medical market compelled numerous Western medical doctors to study this mysterious acupuncture treatment. Western medical insurance companies and governing bodies were subsequently forced to reconsider their policy toward it.

The prevailing attitude of condescension toward ancient treatments experienced a transformation in the 1970s. In many cases, the apparent inability of Western medicine to provide relief from chronic pain and suffering, along with the increased disposable income that the era afforded, led many to try alternative therapies.

Even then, with the exception of a handful of doctors and scientists, acupuncture was largely unknown outside China. This changed in 1976 with the death of Mao Ze-dong, the last absolute ruler of China, as China ended twenty-seven years of isolation to begin to open its doors to the West. Through this narrow portal of communication appeared reports of acupuncture's effectiveness in relieving pain and its use as anesthesia during operations. The first reports about acupuncture appeared in the *New York Times* on April 28, 1971, and it made front-page news on July 25 of that year.[3] In response to this development, doctors in the United States either voiced incredulity or attributed acupuncture's benefits to the placebo effect.[4] Nevertheless, people in the United States and other Western countries became aware of acupuncture, providing it a foothold at the base of Western medicine.

The 1980s saw China open further to the West, and thousands of Western doctors traveled to China to study acupuncture themselves. Ultimately, the situation was comparable to the story introducing this chapter, about the Chinese peasant who used antibiotics from the West to heal without understanding how they worked. Similarly, Western doctors learned how to insert needles into acu-points to benefit their patients without knowing acupuncture's working principles.

Acupuncture became increasingly popular in Western countries. By the beginning of the 1990s, the German Association of Acupuncture already had more than four thousand members. In addition to those studying acupuncture in China, many practitioners had begun to establish courses and open schools to teach acupuncture in Germany and other Western countries.

The collapse of the Soviet Union in 1991 ended the Cold War, which had largely severed communication channels between East and West since the end of World War II. This development led to China further engaging with Western countries, and Chinese medical doctors were permitted to leave China to work in the West. Many headed to the United States, the United Kingdom, and Australia to practice therapies, including acupuncture, herbal medicine, and Qigong.

The twenty-first century has seen clinics offering Chinese medicine achieve an unprecedented level of popularity in the West. In the United States and the United Kingdom, this popularity has begun to make significant inroads into the conventional Western medicine market. This prompted the governments of some Western countries to consider and enact new legislation to regulate the alternative medical market, covering factors such as the qualifications of complementary medical practitioners and the safe application of herbs. Meanwhile, the competitive dynamics of the medical insurance industry have forced companies to adapt their coverage of alternative therapies to satisfy their customers' demands.

Scientific or Not?

It seems likely that these exotic forms of medicine will continue to enjoy rapid worldwide expansion in the medical market. However, practical success does not mean these alternative or complementary forms of medicine have already achieved scientific status. Their success is regarded as purely empirical and without a reliable scientific basis. Despite the fact that a remarkable percentage of acupuncture patients benefit from the treatment, practitioners are unable to explain how their treatment works from a Western medical point of view. This scenario does not sit well with the conservative scientific community, which maintains a cautious, even suspicious, attitude toward these therapies.

For instance, in the last five decades, many anatomists and histologists have searched, without success, for the mysterious acu-points and acu-channels, which are clearly described in the ancient books on acupuncture. The apparent absence of these structures can be seen to invalidate Chinese medical theory and leads to the suggestion that the benefits of acupuncture may be attributable to a mere placebo effect. In other words, the whole chronicle of acupuncture, spanning thousands of years, might only be an elaborate fairy tale, or worse, a colossal hoax.

The increasing popularity of acupuncture and other traditional medicines has led to new research foundations for complementary medicine, including the Foundation for Integrated Medicine in the United Kingdom, supported by the royal family, to be established to encourage serious scientific research. Various private foundations are also conducting research into complementary medicine.

In shaping the administrative and legal framework for the emerging alternative medicine market, government agencies have sought to clarify the situation through scientific research. For example, the National Institutes of Health (NIH), the U.S. government's medical research agency, established the National Center for Complementary and Alternative Medicine in Bethesda, Maryland, to support research. Similarly, government foundations in some Eastern countries, such as China and India, have supported such research for decades.

Universities in the West, even highly conventional institutions such as Harvard Medical School, the University of California, Duke University, the University of Arizona, Imperial College London, the University of Westminster, the University of Greenwich, and Middlesex University, have started to seriously consider complementary medicine. In addition to holding conferences, some have established research teams to investigate the reliability and scientific merit of alternative therapies.

There are two aspects to this research: the first studies the effectiveness, safety, and cost-benefit relationship of various forms of complementary medicine. This research can in some ways be considered a vehicle for skeptics, enabling them to discern whether positive outcomes can be attributed to a placebo effect and whether these alternative systems are genuine therapies or merely the domain of charlatans. This relatively straightforward form of research will provide information concerning the credibility of these therapies, which is highly important for patients and government authorities.

The second aspect is fundamental scientific research into the mechanisms underlying holistic medicine. In some ways, this form of research is for believers, particularly those who want to peer into the mysteries that facilitate alternative therapies. This form of research is not as urgent as the first, but its success would serve two purposes: providing a scientific rationale for complementary medicine, thereby legitimizing its position in society; and enabling major improvements in techniques and effectiveness by employing rational scientific understanding and the power of modern technology.

Revolutions within Science

Fundamental research into the mechanisms of complementary medicine is intertwined with an important move away from the approach of reductionism and the limitations of the particle pattern toward a holistic style of thinking. Such a move would have profound influences not limited to medicine but extending to science, technology, and the future of humanity.

It is possible to draw parallels between the current revival of older therapies and the fifteenth century's Renaissance, which had the outward appearance of a movement directed toward the past, specifically the culture of ancient Greece. In reality it was a progressive movement that marked the beginning of a new epoch that brought about the Reformation, the industrial revolution, modern democracy, modern science, and a new mode of thinking that differed completely from that of the Middle Ages.

Similarly, the current revival of ancient forms of medicine appears to represent a movement or doctrine of returning to the ancients. In reality, it is also part of a progressive movement indicative of the beginning of a new epoch. We are fortunate to live in this period and witness the many transformations that are occurring in medicine, science, technology, society, and our way of thinking.

As stated earlier, this revival of older therapies represents the external level of a major revolution that is discernible to all. Another revolution is currently unfolding at an internal, deep, and fundamental level. Along with the study of the underlying mechanisms of complementary medicine, this revolution will reveal more about the nature of this new epoch. These two revolutions echo and reinforce each other, and both will influence the ultimate form of the new era.

In the last half century, fundamental scientific research has exposed many weaknesses in Western medicine, biology, physiology, psychology, society, and our way of thinking. Let us consider these weaknesses in light of the two evident limitations in the current mode of thinking in medical research.

The Limitation of Purely Chemical Models in Medicine and Biology

As discussed in chapter 2, Stapp pointed out that psychology has moved toward the concepts of nineteenth-century physics while physics has moved in the opposite direction. Nineteenth-century physics employed concepts grounded in materialism, chemicals, and ball-like particles. Physics had moved beyond these incorrect concepts by the early twentieth century, but these obsolete notions, with their inherent materialist limitations, still dominate biology, physiology, and even psychology. The modes of thought overshadowing these disciplines are akin to the outmoded thinking of the Middle Ages. It is impossible to discern the mechanisms behind these ancient forms of medicine, which are related to subtle energy vibrations, information, wave patterns, and harmonies, without changing the way contemporary biology, physiology, and psychology think.

The Limitation of Reductionism

The basic ideas and methods in contemporary medicine, biology, physiology, and psychology are based on the approaches of reductionism, biochemistry, and conquest. Reductionism involves disassembling a system into smaller and smaller separate components in order to study the details of the system, thereby determining which component is responsible when problems occur.

The approach of reductionism was tremendously successful in medical research and the development of biology and psychology in earlier periods. In studies of infectious diseases, scientists would separate bacteria from patients' excretions, passing them through multiple procedures to separate the individual species. At a certain stage it would be determined which species was responsible for the disease, and scientists would then invent some means, such as antibiotics, to kill it.

The approach of reductionism still has some success in the study of genes and genetic diseases today. However, by reducing problems in systems to a single element to be eliminated, little consideration is given to the dynamic interplay of the multitude of elements. Within the approach of reductionism and conquest, there is no trace of the concepts of balance, cooperation, coordination, and harmony, which are fundamental to holistic medicine. It is therefore not possible to study holistic medicine without moving away from the reductionist approach.

Fortunately, a movement away from reductionism toward holism in physics was initiated in the 1970s by modern physicists such as Hermann Haken, Fritjof Capra, Ilya Prigogine, and Ke-hsueh Li. Haken developed the idea that "the whole is bigger than the sum of its parts" under the name synergetics. Capra pointed out that everything in the universe is interconnected in a huge network of strong or weak interactions. Prigogine found the dynamic dissipative structure, and Li proved that the uncertainty principle is not only valid in the microworld but also in the macroworld.[5]

Unfortunately, only a few biologists, physiologists, and psychologists are aware of these major changes in the concepts of physics. For these disciplines to comprehend the essence of holistic thinking, it is necessary, even urgent, to introduce them to these radical concepts. This will provide a framework that allows ancient Eastern therapies to be interpreted from the viewpoint of modern science.

In the following chapters, this important revolution in science will be introduced, alongside an introduction to the history of fundamental scientific research into complementary medicine, focusing particularly on acupuncture and related medicines. Long before the present boom in the complementary medical market, pioneering scientists had already done a lot of exploration. Revolutions in science, particularly in the 1970s and 1980s, have already cleared the way to understand the essence of balance, cooperation, coordination, and harmony, which are fundamental to holistic medicine.

It is worth noting that the outcome of the revolutions in science will not be limited to the establishment of a new united medical system that incorporates Western medicine and complementary medicine under one theoretical system. It will also promote a profound revolution in physics, biology, physiology, psychology, economics, and finally, in modern society's way of thinking.

CHAPTER 5

QUEEN VICTORIA STUDIES TV

Acupuncture is a living and stubborn challenge to established "scientific" knowledge. Its roots are at least four thousand years old, and it is based on a philosophy and view of the body-mind that is entirely different from modern views. It is a total anachronism but it refuses to disappear. If Qi and channels really exist, then modern "scientific" views of the body-mind clearly need to be revised. —GIOVANNI MACIOCIA

Function versus Structure

The natural consequence of the current surge in Eastern medicine is the requirement to pursue modern scientific research into the underlying mechanisms. At first glance, modern scientific basic research into acupuncture would appear to entail seeking an explanation for the workings of the ancient therapy. On closer inspection, however, this pursuit represents a collaborative synthesis between Eastern and Western culture. Modern science and conventional Western medicine were wholly developed in, and based on, Western culture and thinking. As such, this research necessitates attempting to understand a system of medicine derived from Eastern thinking from the viewpoint of Western people and culture.

One of the biggest differences between Eastern and Western culture is that Eastern medicine, in particular acupuncture practice and theory, pays more attention to function than to structure. This inclination stems from a deep-seated principle that as long as a form of medicine works well, it is a good one. Conversely, Western medicine is more focused on structure, based on its entrenched conviction that, in the words of French philosopher René Descartes, "Man is machine"—the human body is merely a complicated machine and medical doctors are engineers. The more the engineers understand the structure of the machine, the better they are able to repair it.

The conviction that the body is a machine provides powerful motivation to disassemble the machine into smaller and smaller parts. Anatomy dissects the human body into various organs, then cellular biology dissects the organs into their various microscopic cells and components, and finally molecular biology dissects cells into their various molecules. As things currently stand, anatomy, cellular biology, and molecular biology have laid down a very solid and reliable structure for modern Western medicine.

If the structure that corresponds to the function of Eastern medicine can be discerned, both Eastern and Western medicine will enjoy a common base. This simple motivation initiated modern scientific basic research into acupuncture after World War II. The journey of discovering the structure that corresponded to the function of acupuncture can be divided into five steps, discussed under the following headings.

Are There Little Puppets inside the TV?

Researching acupuncture is not a simple matter of translating between two languages. The function of acupuncture was not based on the workings of a rationally designed machine, but instead developed entirely from experience and intuitive insight. Difficulties arose when scientists attempted to disassemble "the machine" of the acupuncture system.

Initially, scientists believed they could reveal the mechanics of the acupuncture system as soon as they employed the powerful methods and knowledge systems of anatomy, cellular biology, and molecular biology. Unfortunately, some functions of acupuncture exist beyond the scope of these disciplines, and the research consequently became an extremely difficult task.

To illustrate the early period of this research, from the 1950s to the 1980s, let us explore an analogy where scientists in the time of Queen Victoria attempt to come to terms with the mechanism of television. Suppose the queen was fortunate enough to be given a television set by an alien civilization. Aside from powering itself and tuning to a signal beamed in from the alien home world, this television was identical to those of our everyday experience. It functioned, but the inhabitants of the era, unaware of the concept of electromagnetic fields and waves, did not understand the working principle.

The queen invited a group of prominent scientists to devise a method to investigate the mechanism behind the mysterious playing-box. Their first, very reasonable, supposition would be that there were small puppets acting behind the window of the box. Thus, as soon as they opened the box, they could immediately capture these miniature actors and display them.

The underlying logic of scientists' suppositions at the onset of modern science's investigation of acupuncture was almost identical. Initially, many scientists hypothesized that acupuncture meridians might be a kind of pipeline or cable, like a blood vessel or nerve fiber. In this context, the acupuncture point would be a node on the cable, like a nerve node.

In 1963, North Korean biologist Bonghan Kim announced that he had found a new system composed of "Bonghan corpuscles," which corresponded to acupoints, and "Bonghan ducts," which corresponded to acu-meridians. Cold War dynamics led to his work being prematurely overpropagated by newspapers and magazines in North Korea and the Soviet nations, establishing him as a national hero. Given the potential contribution to the development of medicine, it was also captivating news among the international scientific community.

According to the criteria of modern science, any experiment has to be reproducible by other scientists under the same conditions. If an experiment is reproducible, it is regarded as a true and reliable result. The importance of this discovery led many scientists in China, Germany, France, Austria, and the United States to repeat his method, but no one else could confirm his discovery.

After carefully repeating Kim's experiment, noted Austrian histologist Gottfried Kellner wrote a long paper titled "Structure and Function of Skin." In his opinion, the "Bonghan corpuscle" was actually the end body, the residual stump of a capillary remaining after embryo development. Similarly, the "Bonghan duct" was the corresponding residual capillary. While these are real structures in the skin and other parts of the body, and would react to acupuncture stimulation, they do not possess the function of the acupuncture system. This groundbreaking news became a major scientific scandal that led Kim to commit suicide.

A balanced assessment of the incident would concede that Kim was an important pioneer and a victim of the demanding task of scientifically exploring acupuncture. It is, of course, a sad story, but his mistake served as an important lesson in pursuing this research. Attempting to locate a visible anatomical entity that corresponded to the acupuncture system was natural and logical. Mistakes are common in science, particularly in the exploration of a completely new area. The scandal of Kim's work was a result of premature and overexuberant reporting.

Are Acupuncturists Swindlers and Their Patients Deranged?

The scandal involving Kim's research dealt a substantial blow to acupuncture research, steering it into a dark period. Instead of delving into the mechanisms underlying acupuncture, the question became "Does the acupuncture system exist objectively?" Common opinion in the scientific community at the time held that "neither evidence for its efficacy nor a plausible hypothesis for its action can be advanced."[1] This attitude also implied that the theoretical framework of acupuncture was imaginary, the needling procedures based on fantasy, and any effects on patients purely placebo.[2] The ultimate implication was that the entire history of acupuncture was one of swindling practitioners and deranged patients.

An example illustrating the attitude of scientists at the time occurred during the First National Conference of Acupuncture in China in 1977. Patients who reported a sensation being propagated along a meridian after being stimulated by needles were required to undergo psychological testing to ensure they were not suffering from a mental disorder. Naturally, this suspicious attitude was insulting to acupuncturists and other traditional Chinese medical practitioners. Patients' sensations of propagation along meridians are the most important indication used by acupuncturists to judge the effectiveness of a needling operation. It is untenable that countless numbers of patients had been misleading acupuncturists for thousands of years.

In order to prove that the acupuncture system was not a fantasy, a large-scale investigation was organized in China during the 1970s, focusing on the phenomenon of sensation propagation along meridians. The project marked the first time that the reliability of acupuncture theory was assessed from the viewpoint of modern science, and thus from the viewpoint of Western culture and thinking. It is also the largest research project undertaken in China to date, involving twenty-eight institutions and thousands of medical doctors from all over China, even extending to Chinese medical teams in Africa. In a period spanning six years, 63,228 people of varying gender, age, and race were tested.

In the context of acupuncture, "sensation" refers to the feeling patients experience during needling operation. When successfully directed to the correct position, a special sensation occurs that moves slowly along the corresponding meridian. The feeling is usually experienced as something like sourness, swelling, numbness, pain, warmth, coldness, or an electric shock. Occasionally, one of these sensations occurs in isolation, but normally it is experienced as a mixture of sourness, swelling, and numbness. In terms of the ancient theory of acupuncture, these sensations are attributable to Qi.

In some ways, this form of investigation resembles a public opinion poll. However, even though it is related to subjective experience rather than objective measurement, it complies with modern scientific standards. The methodology of the investigation was also fully standardized in accordance with the criteria of modern science. For example, the methodology stipulated the feeling of Qi stimulated by a low-frequency-pulse electronic instrument using a silver electrode of 0.1 to 0.2 inches in diameter. These electrodes were applied to "well-points," which are on the tops of fingers and toes, or "source-points," usually on the joints of wrists or ankles. The vast majority of scientists in the project conformed to this standard. However, some used needles or hand pressure to stimulate, and some stimulated other acu-points instead of well-points or source-points.

The degree of sensation and the forms to record results were also standardized to enable statistical analysis. The investigation found that 78 percent of the subjects experienced the feeling of sensation propagation along meridians. Obviously it is difficult to assert that over 40,000 people were all deranged in the same manner. As it is established that a purely placebo effect plays, at most, a role in 25 to 30 percent of cases, the result of 78 percent experiencing sensation propagation far exceeds what could be attributable to that.

Asking the Questions Ancient People Didn't

The outcomes of this project extended far beyond the original scope, and many investigators took the opportunity to ask a number of questions not asked in ancient times. For two thousand years China was dominated by Confucianism, which served as both the official national religion and a standard of human behavior in society. According to its principles, it was forbidden to question the word of an authority, let alone to criticize or challenge it.

While very effective in preserving the power of large central governments and totalitarian rulers, Confucianism did not facilitate the development of science. The first questions concerning the veracity of acupuncture were asked by Chinese scientists who were educated in Western science and the principles it derives from Western culture. They took the opportunity to look for answers in their observations, allowing them to clarify many vague concepts in acupunc-

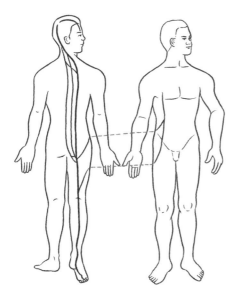

Figure 5.1. Superposition of the routes of sensation propagation along the Bladder Meridian, from one hundred test subjects.

ture. This ultimately helped scientists reveal the secrets behind the mysterious acupuncture system.

Accuracy of the Ancient Records

One of the questions that were forbidden at colleges of Traditional Chinese Medicine was whether the classical texts contained any errors in their description and teaching of acupuncture. The network of the meridian system is clearly depicted in the ancient texts, and the researchers in the project, unfettered by the strictures of Confucianism, took the opportunity to evaluate their accuracy by comparing the routes of sensation propagation with the locations of meridians stipulated in the texts.

Generally speaking, the routes of sensation propagation and the corresponding meridians in the classic texts coincided closely on the limbs. On the trunk, however, they were somewhat shifted, and the difference on the head was quite significant. The routes of sensation propagation also differed among subjects, although in most cases the difference was minor.

Stability and Flexibility of the Route of Sensation Propagation

Interestingly, when a person is seriously ill, the routes of sensation propagation can be completely altered (fig. 5.2). Instead of the original paths, the sensation propagation usually travels directly to the focus of the disease. This easily observ-

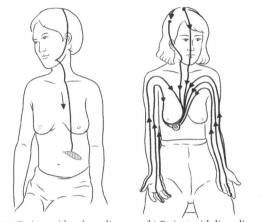

(a) Patient with spleen disease (b) Patient with liver disease

Figure 5.2. Sensation propagation moves to the location of the focus of disease.

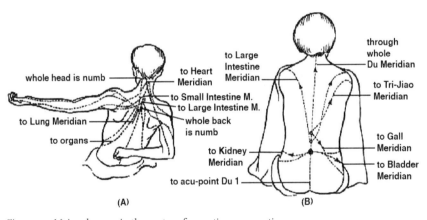

Figure 5.3. Major changes in the routes of sensation propagation.

able phenomenon was recorded in the ancient texts and indicates that a meridian is not a fixed channel, like a blood vessel or a nerve fiber, but rather something variable that adapts to one's physical state. Unfortunately, most scientists searching for the structure of the acupuncture system ignored this important phenomenon.

In rare cases, even larger changes in the path of sensation propagation can be observed (fig. 5.3). Consequently, it is unnecessary to search for a fixed structure that corresponds to the acupuncture meridian network.

Repeat observations, where the same needling operations were performed every day on the same person, were also made during this project. They revealed that, provided experimental conditions are consistent, the routes of sensation propagation are relatively stable, on the order of 86.7 percent, while other cases exhibited 0.4- to 0.8-inch deviations. This implies that the routes of sensation propagation can vary from one day to another, even if the same point is stimulated on the same person.

These observations suggest that the record in the ancient texts is essentially correct, but not perfectly so. Another conclusion that may be inferred is that, contrary to what many once believed or expected, the meridian is not a material entity. These two conclusions were important for the subsequent research into the mechanics of acupuncture.

Width and Depth of the Route of Sensation Propagation

The ancient records of acupuncture contain no information concerning the width and depth of the meridians. In the 1970s research project, however, the width and depth of the route of sensation propagation were observed and recorded.

For most people, the route of sensation propagation is not the thin thread demonstrated in the texts, but rather a band with a central region and a margin (at left in fig. 5.4). The width of the central region is narrow, at 0.08 to 0.2 inches, and is associated with a clear feeling of sensation propagation. The margin is broad, at 0.8 to 2 inches, and is associated with incidents of weaker, vaguer sensation propagation. Another interesting phenomenon is that when the needling location is shifted slightly from the center of the acu-point, the route of sensation propagation also shifts slightly to a parallel position (at right in fig. 5.4).

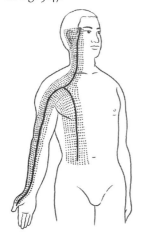
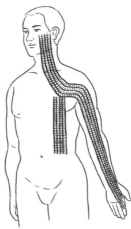

Figure 5.4. The route of sensation propagation along the Lung Meridian.

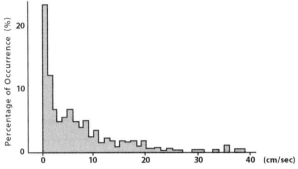
Figure 5.5. Speed distribution of sensation propagation.

Direction and Speed of Sensation Propagation

Unlike blood circulation, sensation propagation usually progresses simultaneously in both directions, proximally (toward the center of the body) and distally (away from the center of the body). At around 0.4 to 8 inches per second (fig. 5.5), the speed of the progression is usually quite slow compared to signals in nerve fibers, which travel at 6 to 400 feet per second. It is worth noting that the speed of signal propagation along meridians later proved to be the key to revealing the mechanism behind the acupuncture system.

Ear-Needling and Sensation Propagation

Many investigators in this research project found that stimulating different acu-points in the ear that corresponded with different meridians could induce several routes of sensation propagation. The sensation propagation initially progresses inside the ear, then across the outer ear and along the related meridian.

Enhancing Sensation Propagation

This project also revealed that sensation propagation could be enhanced by numerous factors: increasing body temperature using a bath; injecting certain drugs such as adenosine triphosphate (ATP), coenzyme A (CoA), and cytochrome c; ingestion of some Chinese herbal remedies; and even meditation.

A Real Event in the Body, or Merely an Illusion in the Brain?

The evident contradiction between the established presence of the function of acupuncture and the absence of a corresponding structure posed a major conundrum for scientists engaged in acupuncture research. The question arose

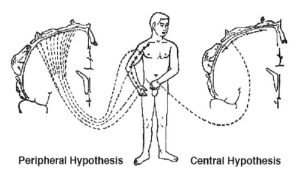

Figure 5.6. Competing hypotheses to explain sensation propagation.

whether the phenomenon of sensation propagation along meridians was a real event in the body or merely an illusory sensation caused by activity in the brain.

There are two major schools of thought in explaining the cause of sensation propagation. These are generally referred to as the central hypothesis and the peripheral hypothesis. The central hypothesis (at right in fig. 5.6) asserts that sensation propagation along meridians exists only in the brain and does not involve the body. On the surface it would appear that the central hypothesis is a variation of the opinion that acupuncture is merely a fairy tale. Adherents concede the objective existence of the acupuncture system but think it is only the activity of the cerebral cortex:

1. The stimulation of an acu-point is transmitted by nerve fiber to a corresponding position in the cerebral cortex.
2. This site in the brain then secretes certain chemical compounds, which diffuse to neighboring locations and trigger a response in those areas.
3. These neighboring locations are connected by nerve fibers to locations on the body that correspond with the route of sensation propagation.
4. Thus, chemical-triggered responses within the brain provide the feeling of sensation propagation.

This is an attractive and elegant hypothesis, and physiologist W. J. S. Krieg produced an impressive diagram to demonstrate it. It divides the skin into twelve regions that are reflected in the cerebral cortex and correspond to the twelve major meridians in the body (fig. 5.7).[3]

As elegant as it is, the central hypotheses must answer some challenges. For instance, inconsistencies in the relationship between the routes of merid-

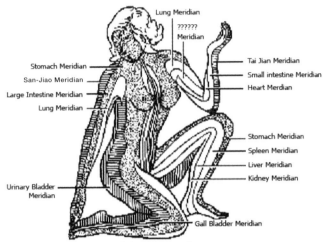

Figure 5.7. Krieg's assumed distribution of somatosensory areas corresponding to the somatic cortex.

ians and their projection in the somatesthetic receptive field of the cerebral cortex must be addressed. It is evident from figure 5.8 that the sensation projection on the somatesthetic receptive field of the cerebral cortex would have to leap over the wide area assigned to feelings from the arm and hand when sensation propagates along any of the three Yang meridians, which run from foot to head.

While numerous cases of sensation propagation can be explained by the mechanism outlined in the central hypothesis, it has more difficulty explaining many other experimental results. These include visible physiological and pathological change along meridians and other measurements outlined below.

On the other hand, there is plenty of experimental evidence to support the peripheral hypothesis (fig. 5.6), which holds that sensation propagation is a process occurring in the body rather than the brain. For example, the correlation between areas of lower resistance and lines of lower resistance on the skin with acu-points and meridians, discovered by Croon and Nakatani, is well established, and is discussed in depth in chapter 6. Vol's subsequent systematic development of this discovery has seen it widely applied in Germany and other countries since 1953 as a standardized diagnostic method.

Also, since the 1926 invention of Kirlian photography, which captures high voltages and high frequencies, many scientists have employed it to detect acu-

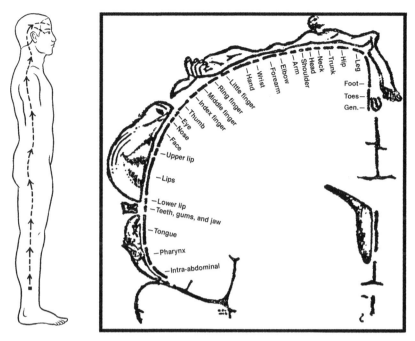

Figure 5.8. Comparison of the three Yang meridians of the foot and the somatesthetic receptive field of the cerebral cortex.

Figure 5.9. High-voltage, high-frequency photograph of a meridian.

points and meridians. Continued improvements in this technique now allow meridians to be clearly demonstrated in a darkened room (fig. 5.9).

Additionally, doctors and medical practitioners using soft laser therapy in the West have found that the meridians coincide with channels of light. Chinese physicist Bi-wu Zhang found that the meridians appear as channels of microwaves. Various other observations reveal that in addition to appearing as channels of electromagnetic waves, meridians are also a channel for acoustic waves (fig. 5.10).

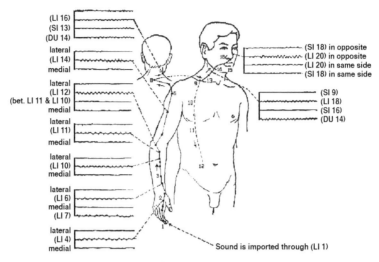

Figure 5.10. Acoustic signals propagation along meridians.

The phenomenon of areas of lower electrical resistance correlating with acoustic channels is not only observable on human skin but also on the skin of animals and plants (plates 5 and 6 in the color plate section).

Finally, isotope tracing techniques, which involve inserting radioactive isotopes into the body and using the radiation emitted to track their movement through the body, show that meridians are not only a channel for waves but are also the path taken by some chemical components. There is no physical channel in the body that can explain why the isotopes follow these paths. Plate 7 in the color plate section shows an isotope following the Kidney Meridian.

Is There Something Beyond Our Present Knowledge?

The contradiction between acupuncture's established function, supported by both clinical success and rigorous scientific observation, and the absence of a corresponding structure is a major challenge to scientists. In terms of chapter 1's analogies, the absence of a physical subject makes the research more difficult than the one facing the blind world's scientists researching the elephant, but similar to their research into the rainbow. This has led many medical doctors and scientists, particularly physiologists, to employ the "ostrich approach." When contemplating acupuncture, they consider only those phenomena that are consistent with current physiological understanding and bury their heads in the sand when confronted with phenomena that contradict this understanding.

Some scientists, sensing the opportunity to extend scientific understanding, paid special attention to these contradictory phenomena. To restate the words of Maciocia that opened this chapter, "Acupuncture is a living and stubborn challenge to established 'scientific' knowledge.... If Qi and channels really exist, then modern 'scientific' views of the body-mind clearly need to be revised."

Third Balance System Hypothesis

In 1983, Chinese physiologist Mon Zhao-Wei asserted that the critical factor of sensation propagation is its speed, usually in the range of 0.8 to 2.8 inches per second. This corresponds neither with the speed of the nervous system nor the speed of bodily fluids. He proposed that the acupuncture system is something new, which he called the third balance system.

According to his framework, outlined below in Table 2.1, the first balance system is the motor nervous system, which controls the movement of voluntary muscles, allowing for dynamic balance in rapid movements, like those occurring in sports and daily work, with a transmission speed of around 230 to 400 feet per second. The second balance system is the autonomic or vegetative nervous system; this regulates the slower dynamic activity of internal organs, such as breathing, digestion, and heartbeat. The third balance system is the meridian system, which transmits stimuli from the surface of the body to the internal organs, allowing for harmonious coordination between these two aspects of the body, thereby maintaining an even slower dynamic balance. The fourth balance system is the endocrine system, which maintains the slowest balance in bodily systems.

Table 2.1. The four balance systems in the body and the speeds of their signals, in feet per second.

Balance System	Speed	Function
first balance system: motor nerves	330 feet/second transmission	rapid posture balance
second balance system: autonomic nervous system	3.3 feet/second transmission	balance of organ activity inside the body
third balance system: meridians	0.3 feet/second sensation propagation	balance between organs and the body's surface
fourth balance system: endocrine system	0.003 feet/second diffusion	slow balance of the whole body

While he did not elaborate details about this new balance system, he was the first physiologist who clearly stated that the mechanism of the acupuncture system could be an unknown new structure. If this new structure was discovered, it would necessitate a completely new chapter in physiology.

The third balance system predicted by Wei is in fact the dissipative structure of electromagnetic fields in living systems, discovered ten years later. To some extent this discovery reconciles the function of the mysterious acupuncture system with its corresponding structure. The discovery of this dissipative structure, which can be thought of as an invisible rainbow and inaudible music inside and around the body, is not limited to opening a new chapter in physiology; it will also facilitate revision of the scientific views of biology, psychology, and medicine.

Wave Guide Channel Hypothesis

On a personal note, I was surprised when my research revealed the existence of another visionary who made a similar prediction long before Wei and Maciocia. In 1959, Chinese physicist Bi-wu Zhang, from Qingdao Medical College, contended that modern medical research focused too heavily on the interactions of materials and not enough on the interactions of energy, that Western medicine overemphasized the importance of solid particles—molecules and atoms—while almost completely ignoring the waves in human bodies and the influence of changing factors such as light, electromagnetic fields, and cosmic rays.

The key point of his wave guide channel hypothesis was that the human body contains many tubelike and sheetlike structures, which are composed of heterogeneous media. These media interact with visible light in differing ways. Stated scientifically, they possess varying reflection coefficients, refraction coefficients, and polarization spectra toward visible light. These varying properties allow us to study these structures with the naked eye and the microscope. It can be assumed that these structures also vary in their interactions with infrared rays and microwaves, which are plentiful in the body. Therefore, the structures formed by these different materials interact to create a wave guidance system that directs the transmission of electromagnetic waves inside a body.

Zhang regarded the concept of internal Qi in the theory of Traditional Chinese Medicine to be electromagnetic waves within the human body and named these electromagnetic waves Qi-photons. He speculated that:

Figure 5.14. Chinese physicist Bi-wu Zhang (1920–1992), the first person to point out the important role of electromagnetic waves in the acupuncture system and other holistic forms of medicine.

- Qi-photons in the body had the same importance as materials—molecules and atoms.
- The relationship between different meridians, between meridians and their acu-points, and between meridians and corresponding organs could be explained in terms of a wave guidance system.
- The line with the highest probability of finding Qi-photons would be the axis of a meridian, around which there may be a tubelike distribution of Qi-photons that form its boundary.
- The slow speed of Qi or sensation propagation along meridians could be attributed to the "group speed" of the traveling waves as they progressed along their path.

I first encountered Zhang's work in China in 1994, after I had developed the idea of the electromagnetic structure in the body in Germany in 1992. I was stunned by the visionary brilliance of his hypothesis. He had proposed almost the entire mechanism of fields and waves within the human body. The only aspect that he had overlooked was the relatively minor step of considering dissipative structures of electromagnetic fields, although labeling this omission an oversight is unfair. The theory of dissipative structures was only developed in the 1970s by Belgian scientist and Nobel laureate Ilya Prigogine. The integration of the role of dissipative structures into an empirical analysis of the human body depended on subsequent mathematical advances such as chaos theory, fractals, and many other nonlinear mathematical problems, discussed in chapter 7.

CHAPTER 6

BLIND SCIENTISTS DISCOVER THE RAINBOW

> A fish said to another fish, "Above this sea of ours there is another sea, with creatures swimming in it—and they live there, even as we live here." The fish replied, "Pure fancy! When you know that everything that leaves our sea by even an inch, and stays out of it, dies. What proof have you of other lives in other seas?"
> —KAHLIL GIBRAN, "Other Seas," *The Forerunner, His Parables and Poems*

Accumulated scientific research and observations over the last half-century, coupled with developments in physics over the last three decades, have enabled the veil obscuring the mysterious mechanism of acupuncture and related holistic medicines to be incrementally drawn away. The development of acupuncture research can be categorized into four general stages:

1. Anatomical study tried to find anatomical structures that corresponded to the function of acupuncture; the result was negative.
2. Phenomenological study asked whether acupuncture phenomena, such as sensation propagation, low resistance, and other physiological reactions, actually existed; the result was positive.
3. Physiological study attempted to explain acupuncture phenomena within the existing knowledge of physiology and was able to propose feasible mechanisms for some of the phenomena. When it came to established phenomena that could not be explained by the existing framework, however, the approach was to pretend they didn't exist.
4. Frontier study focuses specifically on the phenomena that challenge the framework of existing knowledge, with the intent of extending

our understanding. If successful, this research will make important contributions not only to medicine but also to physiology, biology, psychology, and even physics.

Asking Naive Questions

According to the principles of Confucianism, a respectable individual should avoid asking questions that would embarrass or displeasure authority. The essence of frontier science, however, is perpetually asking questions, including naive ones, and tirelessly pursuing the truth behind phenomena. The following sections are some of a long list of ostensibly ignorant questions that could be posed concerning acupuncture.

How Big Is an Acu-Point?

This straightforward question elicits a relatively vast range of responses from acupuncturists. Some say that it is in the size of a sesame seed, others claim it is the size of a soybean, and others believe it is even bigger. The diversity of responses encouraged me to make posing this question to acupuncturists, whenever the opportunity arose, something of a hobby.

The most humorous answer came from an experienced Chinese acupuncturist who was about fifty years old. With a friendly smile, he responded, "I asked the same question when I was a student. But my teacher got angry with me and scolded me so awfully that I never dared to ask such a stupid question again." This is a typical Confucian response, admonishing the student for the impudence of posing a question that might embarrass his teacher.

A German doctor provided a completely different answer. Lacking the Confucian influence and being educated in Western medicine, with its firm belief that the body is a machine, he responded without hesitation, "Oh, the acupuncture point is round and 2.5 millimeters [0.1 inches] in diameter." I subsequently discovered that the head of the electrode in the electro-acupuncture device he used was flat, round, and 0.1 inches in diameter.

As discussed, people long believed that meridians were some form of physical channel, like a blood vessel or nerve fiber, and acu-points were akin to ganglions (clusters of nerve cells) or some form of hole, but anatomy and histology could not prove it. The existence of acu-points and meridians has been objectively proven by means of electronic measurement of skin conductivity with excellent

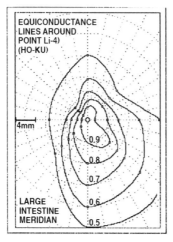 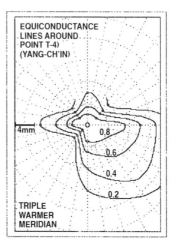

Figure 6.1. The shape of acu-points.

reproducibility, and so electronic measurement allows a relatively straightforward path of inquiry into the form of meridians and acu-points.

The images in figure 6.1 are derived from measurements made by the American scientist R. O. Becker at New York University in 1960. They look like contour maps of hills, but rather than lines of equal altitude, they show lines of equal electrical conductance, which is the ease with which an electrical current passes. These images illustrate why acupuncturists find it so difficult to answer the question "How big is an acu-point?" with any consistency. It is comparable to asking, "How big is the top of a hill?" The answer depends on an arbitrary definition—which contour line you select on a map. So the answer to the first naive question is that acu-points, unlike nerve knots or holes, have no clear boundaries. Instead, they may be considered as akin to small invisible hills with subjective boundaries.

The technique of electronic measurement also allows the shape of meridians to be explored. Figure 6.2 depicts the results of electronic measurements made along the meridian (left) and across the meridian (right), by American scientist R. O. Backer in 1960 and Chinese scientist R. J. Zhang in 1980, respectively.

The results show that the form of a meridian can be visualized as somewhat like a miniature invisible mountain range, and as such do not possess clear boundaries. Several peaks, coinciding with acu-point locations, occur along this range. The research done in China in the 1970s into the width and depth of routes of sensation propagation, discussed in chapter 5, is consistent with this conclusion.

Blind Scientists Discover the Rainbow

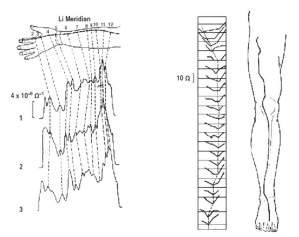

Figure 6.2. The shape of meridians.

In 1986, W. P. Zhang, a young physicist specializing in hydraulics at the Chinese Academy of Traditional Chinese Medicine, studied a low-resistance liquid channel beneath the skin with a hydraulic instrument. He found that this channel, to some extent, also corresponds with meridians, and that its shape is similar to the route of sensation propagation along meridians, discussed in chapter 5: that is, the channel is like a band with a central thread and two margin regions. The width of the central thread is narrow, while the margin regions are broad with nebulous boundaries. The experiments described in chapter 5 about tracing the path of radioactive isotopes through the body also support this finding.

The combined implications of the conclusions of these multiple studies of the shapes of acu-points and meridians support the unified conclusion that a meridian resembles a small, invisible mountain range with many small peaks, which we call acu-points. This is an important step toward revealing the reality of the invisible rainbow in our bodies.

Do Acu-Points and Acu-Meridians Move?

Be careful about posing this question to acupuncturists. In my experience, many consider it to be ignorant and respond with derision. At best, you might be directed to any number of elementary textbooks that depict the established routes of meridians and locations of acu-points.

Fortunately, I have also encountered some experienced acupuncturists, such as Ding-zhong Li at the Sixth Hospital of Beijing and Klaus Peter Schlebusch in

Essen, Germany, who appreciated the question and responded that acu-points are not fixed. While they usually have a definite location, with slight variations, they can move, sometimes dramatically. This is particularly true of acu-points on the limbs, which in special cases can exhibit several inches of movement.

These doctors assert that the textbook representation of the fixed network of the acupuncture system is a simplified depiction for educating students. In reality, the meridians and acu-points are, in their words, "vital." When asked how they could find the precise location of acu-points if they have moved, they responded: "By feeling and intuition." This poses a challenge to those of us who, possessing neither sufficient sensitivity nor intuition, have to depend on instruments.

Fortunately, as with inquiring into the form of acu-points, this kind of measurement by instruments is also relatively straightforward. Furthermore, the measurement data can be depicted as images with pseudo-colors, allowing for easier interpretation of the results. Plates 8 and 9 in the color plate section shows that the acu-points on the top of the fingers are relatively stable with only minor variations, while the acu-points on the palms can move significantly.

The extensive investigation into sensation propagation along meridians in China in the 1970s supports a similar conclusion. As discussed, Ding-zhong Li and others observed significant deviations from the routes of meridians displayed in the textbooks, as shown in figures 5.2 and 5.3, meaning the answer to the question is yes, acu-points and meridians do move. This suggests that it is impossible to locate a system of fixed pipelines and knots that correspond with the acupuncture system. As such, anatomical research into the acupuncture system can never succeed.

Are Electronic Measurements on the Skin Reliable?

Scientists generally place much more faith in instruments than in human responses, as instruments do not lie. But instruments are operated by people, who can make mistakes. Consequently the results of measurements by instruments also have to be verified by others.

Consider the example of electro-acupuncture, developed by Voll as an electronic diagnosis system in 1953 and still widely used by thousands of practitioners in Germany. During a conversation with some fellow scientists I was informed of research undertaken at Germany's Kaiserslautern University of Technology that led to findings critical of Voll's system. While I was assured

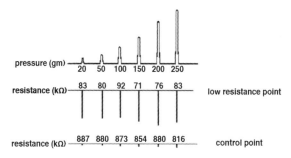

Figure 6.4. Skin resistance and electrode pressure at acu-points and non-acu-points.

that the research had been performed without any bias against acupuncture, the findings pointed to two significant issues with skin resistance measurement. First, the pressure of the electrodes on skin greatly influenced the reading on the resistance meter. The more pressure was applied, the lower the resistance that was recorded. Second, there was a persistent fluctuation in the reading, and as the sensitivity of the recording instrument increased, the stability of the reading decreased. Consequently, it was concluded that Voll's entire system of electro-acupuncture was unreliable.

Fortunately, the same question had already been asked and systematically investigated by numerous Chinese physicists in the 1970s during the expansive sensation propagation research project. The experimental results (fig. 6.4) of Shi Yi, one of the physicists involved, show that the difference between the readings of skin resistance at acu-points and at non-acu-point locations is usually in excess of an order of magnitude—the readings for electrical resistance at non-acu-point locations was more than ten times higher than the resistance at acu-point locations.[1] While the pressure applied to the electrode influences the absolute reading at both types of points, the relative difference is almost independent of pressure. Therefore, while measurement accuracy is not particularly good when assessed under the exacting standards of electronics, it is good enough to locate acu-points during clinical practice. As such, it can be asserted that skin resistance measurement, which forms the basis of Voll's electro-acupuncture, is reliable.

In addition to the relationship between measurements and electrode pressure, Chinese scientists also studied the influence of the voltage used during measurements. Figure 6.5 shows the results of an experiment conducted by

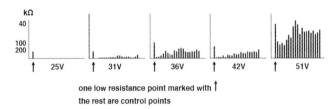

one low resistance point marked with ↑
the rest are control points

Figure 6.5. Skin resistance and measurement voltage at acu-points and non-acu-points. Acu-point measurements are marked with arrows.

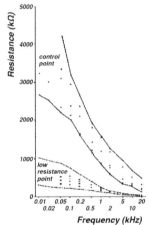

Figure 6.6. Skin resistance and frequency of the measuring current.

Xiang-long Hu at the Fujian Institute of Traditional Chinese Medicine. It is evident that a significant relative difference between measurements taken at acu-points and non-acu-points is maintained for a range of different measurement voltages. While the absolute readings are greatly influenced by voltage, the lower voltages are all able to reliably discern acu-points.

Finally, the influence of electrical frequency on measurements was systematically investigated in 1960 by a young German physicist, C.-E. Overhof, at the Karlsruhe Institute of Technology. This study formed his doctoral research and was performed under the supervision of Croon, the first person to discover the unusual electrical characteristics of the skin at acu-points, as well as W. Ernsthausen and H. Rothe. As with the other investigations, the results of this study (fig. 6.6) show that while the measurement frequency greatly influences the absolute values, a significant relative difference between the low-resistance acu-points and the non-acu-point control points is always present.[2]

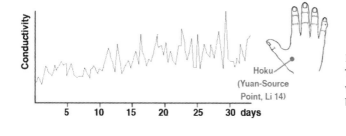

Figure 6.7. The macroscopic wave trains of body conductivity.

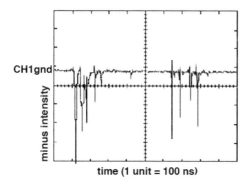

Figure 6.8. Microscopic wave trains in a living system. Image by Konstantin Zioutas, physicist at CERN/LAA.

Are Electronic Measurements on the Skin Stable?

The second issue raised by the researcher at Kaiserslautern University was that in addition to being influenced by pressure, the value of electronic measurement also fluctuated considerably. In addition to being strongly influenced by pathological and unusual physiological states, some spontaneous fluctuation in healthy individuals (fig. 6.7) is consistent with earlier findings. It also correlates, to some degree, with biorhythms and is referred to in the ancient acupuncture texts by the somewhat esoteric term "midnight-noon ebb-flow."

Furthermore, the more precise the instrument used, the larger the fluctuations found in electronic measurements on the skin. An instrument with very high sensitivity to skin resistance fluctuates violently at very high frequencies (fig. 6.8). Measurements of skin resistance exhibit both slow cycles of fluctuation, similar to solar and lunar periods, and very rapid cycles of fluctuation, in milliseconds and even microseconds. From the viewpoint of electronics, skin would be considered a very poor and unstable resistor.

Fortunately, the relationship between the increasing sensitivity of instruments and measurement instability also implies that less sensitive instruments will exhibit much smaller fluctuations. In practice, medical doctors use much less sensitive equipment than those used at Kaiserslautern University, and even less sensitive when compared to the equipment used by Greek physicist Konstantin Zioutas at the European Organization for Nuclear Research in Geneva.[3] The measurement data from clinical instruments is stable enough for clinical practice.

"Skin Resistance Measurement" Is a Misnomer

Most people involved in practice and research into electronic acupuncture use the term *skin resistance measurement* for a number of reasons: the measurements are performed on the skin, and also because the method involved is almost identical to measuring resistance in electronics, implying that these measurements are also of resistance. Finally, within the context of clinical instruments, the resistance of bodily fluids is small enough to be negligible, so readings can only be attributed to the skin.

However, the term is a misnomer that has misdirected research and impeded our understanding of the real mechanism at play in the variations in skin resistance. Rigorous examination of the methodology behind the term reveals inconsistencies with existing knowledge of anatomy, histology, physiology, biochemistry, and physics. Moving beyond the term *skin resistance measurement* lays the foundation for discerning the truth behind these issues.

No Anatomic Evidence—Skin Plays No Role

It has been established that the difference in readings between acu-points and control points is quite significant, usually in excess of an order of magnitude. If such a difference could be attributed to differences in skin structure, it would be relatively easy to detect these by way of anatomy or histology. There is no evidence to support this.

Bodily Fluids Play No Role

Once the possibility of skin producing the variations in resistance measurements is discarded, the search for another source of resistance in the body ensues. As stated above, the conductivity of liquids beneath the skin—tissue fluids, lymph, and blood—is similar to the conductivity of seawater, its resistance is so low as to be beyond the measurement capability of clinical instruments. In other words, the resistance of bodily fluids is so small that it is negligible in these measurements.

Neither Nerves nor Blood Capillaries Play a Role

Another proposed explanation for the variations in skin resistance is that they stem from activity of the nervous system or blood capillaries. However, both the nervous system and the circulatory system are completely immersed in bodily fluids, whose resistance is so small that the activities of the nervous system and capillaries contribute essentially nothing to the measurements.

Consider the following analogy: a high-speed multilane highway (low resistance) and a very low-speed, unpaved, bumpy rural road (high resistance) both run between two cities. The roughness of the rural road will have a negligible effect on people's commute times between the cities because everyone will be driving on the highway.

Acupuncture in Plants, Which Have No Nerves

The case for the nervous system contributing significantly to variations in skin resistance becomes even weaker when we consider that, in addition to being present on human and animal skin, these low-resistance points also occur on plants, as shown in figure 5.12. Scientists at the Xingjiang Forest Institute in China and the Hungarian Biophysics Institute measured the resistance on the bark of trees and defined the lowest resistance points as acu-points.[4] They then inserted needles in these points and monitored the trees with infrared cameras. After 10 minutes, the temperature of the trees increased by 0.5 to 0.7 degrees Fahrenheit, and after two weeks, the shoots of the acupunctured trees had grown more than the control group. As plants have no nervous system, the apparent existence of an acupuncture system in plants contradicts a link to the nervous system.

Sweat Plays No Role in the Lie Detector

The generally accepted explanation for the working mechanism behind lie detectors, which also make use of skin resistance measurements, is also questionable. Mainstream opinion holds that the variation in lie detector readings is attributable to the changes that occur in the sweat on the skin when a subject lies. If this explanation were correct, skin resistance should decrease in a linear fashion as the lying individual starts to perspire. In reality, the readings of a lie detector fluctuate wildly when a lie is detected, meaning that the body would have to be able to repeatedly secrete and absorb sweat at a very high frequency, which is not possible.

Peeling Off the Skin and the Holographic Phenomenon

Japanese physiologist and psychologist Hiroshi Motoyama performed an experiment where he peeled off the *stratum corneum*, the outermost layer of dead skin that forms a barrier to protect the underlying live tissue, finding that only 30 percent of the readings of electronic measurements could be attributed to the outermost layer, and thus 70 percent must originate somewhere beneath.[5] As the layer of dead skin constitutes most skin resistance, Motoyama's results demonstrate that skin resistance is merely background noise, while the bulk of the signal—the 70 percent—originates inside the body.

If skin is eliminated as the major source of variation, where is the real signal coming from? Motoyama suggested that the bulk of the signal could be attributed to the polarization of tissue beneath the skin, near the measuring electrode. In this context, polarization means that the previously electrically neutral tissue takes on positively and negatively charged regions as a result of the application of the charged electrode. This would serve to impede the flow of electricity and increase the resistance of the tissue.

When considered in isolation, the polarization explanation has merit. However, polarization is generally a highly localized event induced by the measuring electrode. In contrast, when a patient is ill, electrical measurements are not only synchronously altered on all main corresponding acu-points throughout the meridian network, but also on all micro acu-points, including those on the ears, nose, palms, and feet. In other words, the change in these electrical measurements is delocalized and holographic—meaning that the state of each body part is reflected in the whole body.

This holographic change in electrical measurements occurs not only at acu-points but also at any point on the skin. That is, whenever there is some change in the body-mind system, the shapes of the probability distributions of the electronic measurement data synchronously change, maintaining similarity in patterns at different locations and on different measurement scales (Plate 10 in the color plate section). This phenomenon is called statistical self-similarity.

The Channel of Light, Microwaves, Acoustic Waves, and Isotope Tracers

Aside from these issues, there are many other phenomena in acupuncture that cannot be reconciled with the present knowledge of physiology. For instance, as discussed in chapter 5, numerous experimental findings illustrate that meridians act as a channel for light, microwaves, acoustic waves, and even isotope tracers. Serious investigation into acupuncture should not avoid the challenge of integrating these phenomena into an explanation of the working mechanism. Instead, the challenge of extending our knowledge to find the answer behind all these puzzling phenomena should be embraced.

Examining Underlying Assumptions: What Is Resistance?

Herein lies the decisive step that enables a breakthrough. Again, it involves posing a seemingly uninformed question: "What does a resistance meter measure?" Any student studying electronics can state the obvious answer without hesitation: "Resistance meters measure electrical resistance," and move on to the next topic. It turns out that this seemingly inexpert question is appropriate and important, while the apparently obvious answer is incorrect and misleading. The following explanation is somewhat technical and mathematical but is central to revealing the underlying mechanism behind the acupuncture phenomena.

A resistance meter measures the electrical current passing through the tested object, not the resistance itself. The value that appears on the meter is actually derived from a purely mathematical calculation that involves applying Ohm's law (resistance = voltage/current, or to use symbols, $R=V/I$). The measurement voltage is set to a known value on the meter itself, and the current in the tested object then creates a magnetic field that affects a coil in the meter, which moves the meter's needle. Consequently, the "resistance" measurements that all the previously discussed studies refer to are actually "current" measurements. For a given voltage, this resistance is inversely proportional to the current. This means that if the resistance is doubled, the current is halved; if the resistance is quadrupled, the current is quartered, and so on.

The conductivity (J) of a substance is a measure of how easily electrons can pass through it. Resistance, which measures how hard it is for electrons to pass through a substance, is thus the inverse of conductivity. Mathematically this can be expressed as $R=1/J$.

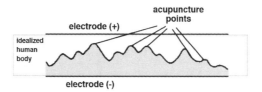

Figure 6.10. The background of electronic measurement.

Conductivity is also proportional to the electrical field strength (E). This can be expressed mathematically as $J = \sigma E$. In this equation σ is a constant, a property of the type of substance in question. Therefore, what is measured on the skin is actually the conductivity for the measurement current, which is proportional to the electrical field inside the body.

Energy Distribution inside the Body

To summarize, the key conclusion of the preceding discussion is that the measurements being referred to as "skin resistance measurements" are in fact measurements of conductivity, which is proportional to the strength of the electrical field inside the body. The following consideration should make the implication of this conclusion clearer. Consider an extremely simplified example of an idealized human body that is cuboid in shape (fig. 6.10). In this situation the electrodes can be regarded as two large flat plates that make contact with the body on both sides; this is consistent with using a constant measurement voltage.

If the diagram in figure 6.10 is compared to the measurement results in figures 6.1 and 6.2, it is evident that meridians are lines of higher-strength electrical field, while acu-points are points where the electrical field is the strongest. This highly simplified diagram illustrates that what is measured on the skin is, in fact, the heterogeneous distribution of electrical field inside the body. In other words, what is measured on the skin of the body is actually the energy distribution inside the body.

A New Chapter in Physiology

The conclusion that skin resistance measurements actually measure the energy distribution within the body has far-reaching significance. Given the significant correlation between the acupuncture system and the electronic measurements, what the creators of this system, using intuitive insight in ancient times, actually discerned is an approximate depiction of the energy distribution, or energy structure, inside the human body. As discussed in part 1, energy is an invisible, untouchable, and ethereal entity that can be compared to an invisible rainbow and inaudible music that exists in humans and other living creatures.

This discovery, elaborated in the following chapters, solves many of the conundrums posed by the puzzling phenomena of the acupuncture system, including:

- higher conductivity at acu-points and meridians
- the slow speed of sensation propagation along meridians
- wild fluctuations in body conductivity
- holographic changes in body conductivity
- meridians acting as light, microwave, and acoustic channels
- the lower resistance channel of meridians
- the isotope channel of meridians

This invisible energy structure can be considered Mon-zhao Wei's third balance system, outlined in chapter 5, and as such opens a new chapter in physiology. In physics terms, this energy structure is a dynamic dissipative structure composed of chaotic electromagnetic standing waves. This development introduces some new concepts that are currently beyond the knowledge base of current physiology, biology, and medicine, and so are unfamiliar to physiologists and medical doctors. In fact, having been developed since the 1970s, they are even relatively new to physicists.

These new concepts, introduced in the following chapters, enabled scientists to discover the beautiful invisible rainbow in our bodies. The recognition for this discovery belongs to the accumulated endeavors of many scientists over several generations. This discovery was as elusive for our scientists as the visible rainbow would be for scientists in chapter 1's world of the blind.

PART 3

DEVELOPING THE CONCEPT OF STRUCTURE

CHAPTER 7

A NEW CONTINENT IN SCIENCE: THE DISSIPATIVE STRUCTURE

Christopher Columbus threw himself down on his knees and thanked God, then stood up and announced the discovery of a new route to India. Everyone appreciates the importance of the discovery of the new continent. For those in Europe, Asia, and Africa, this historic discovery significantly expanded the known world and enabled the birth of new nations. Similarly, discerning the structure that corresponds to the function of acupuncture required the discovery of a new realm in science, namely, the dissipative structure.

Dissipative Structure Occurs Everywhere

While the term *dissipative structure* is academic and unfamiliar to most people, the phenomena it describes exist everywhere. Despite being unaware of the underlying scientific theory, we have all encountered numerous examples of dissipative structures. A waterfall is a typical example: it can only exist when there is permanent and continuous flow of water from an elevated position. In other words, it requires a supply of water with higher potential energy. The waterfall permanently dissipates this energy and can be called a dissipative structure. The flame of a candle is another example. It can exist only when there is a continuous supply of energy—in other words, it perpetually dissipates energy. The natural spring, the artificial fountain, the whirlpool in a river, the terrible hurricane, and the beautiful clouds in the sky are all dissipative structures. Lightning is another example; it dissipates its store of potential energy so quickly that it only survives for a brief moment.

Dissipative structures exist in contrast to the more common "static" category of structures, examples of which include buildings, mountains, and trees as well as moving cars, trains, and rockets. It is worth noting the car, the train, and the rocket are static structures, even though they can be fast-moving and consume energy. The distinction is that these vehicles can be stored while isolated from a source of energy. Conversely, isolation is fatal for any dissipative structure. A waterfall will immediately disappear if isolated from its upriver water supply. In essence, the dissipative structure is vital, while the static structure is dead. This may seem a somewhat simple and self-evident truth, but it took more than one hundred years for scientific theory to come to terms with it.

The Beautiful Dream of Perpetual Motion

The beginning of the industrial revolution brought newfound awareness of the importance of energy. Talented scientists and inventors pursued the dream of building a machine that could run forever, sustained by its own power alone, not requiring any external energy. It became fashionable for aspiring and respected thinkers at the time to consider how to invent such a machine. Even in modern times this is a popular endeavor; for example, the U.S. Patent and Trademark Office received 252 patent applications involving perpetual motion between March and September 2001.[1]

While most of these inventions are extremely ingenious, none of them have satisfied the requirements of running perpetually without access to external energy. After many failures of perpetual motion, scientists finally deduced the law of conservation of energy, which, to date, nothing is able to violate.

The Compromised Dream of Perpetual Motion of the Second Kind

Once the law of conservation of energy was acknowledged, people tried to invent what became known as a perpetual motion machine of the second kind. They theorized that if a machine could be designed to spontaneously convert an abundant source of low-temperature thermal energy into movement, it could essentially operate perpetually without violating the law of conservation of energy. For example, if a machine could be constructed that harnessed the thermal energy in the ocean, this energy could be converted into mechanical work. To put the abundance of this energy source in context, the energy in half a degree in temperature of thermal energy from the ocean could power all the machines that currently exist in the world for about three thousand years. In a practical sense, this would enable perpetual motion.

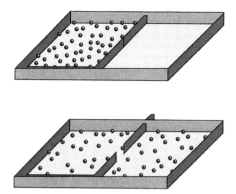

Figure 7.1. An irreversible increase of entropy.

Even the compromised dream of perpetual motion of the second kind is unachievable. After many failed attempts, scientists deduced another important law in physics, the second law of thermodynamics. Perhaps the most understandable expression of this law is, "Heat cannot be transferred from a heat source at lower temperature to a heat source at higher temperature without external energy input." As this transfer of heat energy from a higher temperature area to a lower temperature area is required to convert thermal energy to mechanical work, this expression clearly indicates the failure of perpetual motion of the second kind.

An academic way to express the second law of thermodynamics is, "In an isolated system, entropy irreversibly increases." The first important concept is an *isolated system*, which exchanges neither matter nor energy with its surroundings. Isolating the system in question from its surroundings has been a fundamental, conventional, and routine method in scientific investigation for a long time. Isolation allows the conditions and variations within a system to be clarified and prevents external disturbances. The second concept is *entropy*, which is a measure of the degree of disorder. Quantifying a variable as nebulous as disorder appears counterintuitive, but mathematicians have successfully been able to do so with clarity. The following example of the degree of disorder in a simple system illustrates how entropy can be quantified and why entropy always increases irreversibly in an isolated system. Figure 7.1 shows a plate with 50 small balls on it. The plate is enclosed by a wall, and a partition divides the plate into two sections.

Initially the balls are limited to moving in the left section, even when the plate is continuously shaken (top image in fig. 7.1). If a gap is made in the partition, over time many balls—approximately half—will move from the left section to the right (bottom in fig. 7.1). In terms of the degree of disorder, it can be seen that there is more disorder in the second scenario.

Once the situation of higher disorder has been established, is it possible for random shaking to return all the balls to the left section? Mathematically, it is possible, but with an extremely low probability. The probability is 1 in 2^{50}, or about 0.00000000000000089. This means we might have a reasonable chance of having all the balls returning to the left part after shaking the plate every second, day and night, for 35,702,052 years. In a practical sense, this is impossible. If more balls are involved in the system, the possibility of achieving this declines exponentially. This is the mathematical basis for the concept of an irreversible process, or an irreversible increase in entropy in an isolated system.

Ilya Prigogine, the Modern Columbus

One hundred years after the discovery of the disorder-destined second law of thermodynamics, and just as science's journey of discovery appeared to be approaching its end, a development occurred that moved in the opposite direction—toward order. In the 1970s, the theories of quantum physics and molecular biology were close to perfection, or in other words, close to the end of their spectacular development. Physicists were considering how to unify the four fundamental interactions—gravity, electromagnetism, the strong nuclear force, and the weak nuclear force—in order to establish a final unified theory, or Theory of Everything, that would be able to explain and calculate every phenomenon in the universe, from the most minute quantum aspects to the whole universe itself.

Once achieved, no major questions would remain for theoretical physicists to ponder. Having no major mysteries to explore, theoretical physics research departments could be wound down and, given the existence of a complete and perfect theoretical framework, the exploration of physics theory could be abandoned. Coming generations would be left to study the theory, research its application in niche areas, and search for new technological applications.

Meanwhile, 1970s biology also seemed close to perfection. After the great success of discerning the genetic code in DNA and the revelation of the three-dimensional structure of protein, biologists were considering how to study every molecule in living systems. It would be a massive project but, in principle, would not pose any problems. Once this colossal undertaking was completed, the departments of biology in universities could also be wound down, or perhaps adopt new names, such as the department of genetic engineering or department of biomolecule engineering, as many universities have done in contemporary times. In other words, new ideas in modern science were dying out in the 1970s.

At that point, a modern-day Columbus, Ilya Prigogine (fig. 7.2), made several revolutionary advances in science. He can be regarded as having discovered a new continent of scientific exploration.

Figure 7.2. Ilya Prigogine (1917–2003), the modern Columbus.

From Isolated System to Open System

The first daring step taken by Prigogine was to study open systems instead of isolated systems. In contrast to an isolated system, an open system has an intimate relationship and constant exchange with its surroundings. This includes an exchange of matter, energy, information, and entropy, which can be both positive and negative. In other words, it is possible to import negative entropy into a system in order to reduce the entropy inside the system.

The term *negative entropy* is academic, but it is essentially a simple concept. In the example in figure 7.1, touching the balls on the plate is prohibited, because it is an isolated system. Under this restriction, we can only witness the irreversible increase of entropy. In other words, we can only witness the "decay" of the system and are powerless to reverse it. In an open system, this restriction is lifted, allowing us to pick up the balls in the right section and put them back in the left. This simple procedure reduces the entropy in the system. Expressed in academic terms, we have imported some negative entropy into the system.

This simple idea represented a decisive step. It is illustrated by a story, almost certainly apocryphal, involving Columbus. Upon his return from his historic voyage, some asserted that it was a simple achievement—all that was required was to head west instead of east; anyone could have done it. Instead of arguing, Columbus asked whether anybody could place an egg vertically on the dinner table. When no one was able to, Columbus took the egg and struck it gently on the table. The shell was broken, but it stood vertically on the table. Finding the new continent was an essentially uncomplicated task, but nobody else had been able to do it. This illustrates why it was so difficult for countless intelligent scientists to venture beyond the old continent of the isolated system to the new one of the open system.

The concept of an open system has been accepted and is taken for granted by the scientific community. However, this new realm contains countless unknown phenomena that remain unexplored, particularly in relation to living systems. It is well established that living systems are open systems. As early as 1944, the iconic scientist Erwin Schrödinger (1887–1961) had already pointed out that what we eat and breathe is actually a kind of negative entropy, which allows a high degree of order to be maintained in our bodies. Unfortunately, research into this aspect of biology is still desperately needed. Most biologists are unfamiliar with the terms *entropy, open system,* and *dissipative structure.*

From Equilibrium State to Far from Equilibrium

The second brave step taken by Prigogine was moving from considering an equilibrium state to considering far from equilibrium state. A system categorized as being in an equilibrium state has already reached maximum entropy. This means that it is in a perfectly homogeneous state and exhibits no internal variation.

Figure 7.3 is a cross-sectional view of cooking pots resting on an electric stove. The water in the pot on the left, sitting for some time with the stove off, exhibits homogeneous composition. As such, it is in an equilibrium state.

If the stove is turned to a low temperature, the pot of water will change to the situation shown in the middle pot. The higher temperature at the bottom will gradually and smoothly diffuse to the surface. The heat energy from the stove has shifted the water only slightly from its equilibrium state; this state is referred to as quasi-equilibrium. It still allows for the calculation of entropy, temperature, and other parameters by visualizing the water as many thin slices and regarding each slice as a system in an equilibrium state. This was how experts in thermodynamics dealt with this situation before Prigogine.

However, if the temperature of the stove is increased significantly, the water will boil, as in the pot on the right. Boiling water is in a completely chaotic and turbulent state. Prior to Prigogine, scientists believed that there was no law that could find order in this situation, but Prigogine found that this turbulent state is not as exceedingly chaotic as once believed. Some dynamic structure exists in the chaos, particularly when the energy input is stable.

When viewed from above, boiling water reveals a pattern on the surface of the water that remains stable while the energy supply is stable. These stable patterns

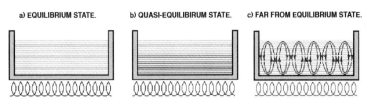

Figure 7.3. From equilibrium to far from equilibrium.

illustrate the essential idea behind Prigogine's discovery. In generalized terms, he deduced that a new order would spontaneously arise from disorder. In other words, new structures will occur in a far from equilibrium state.

From Static Structure to Dissipative Structure

Prigogine named these new structures dissipative structures. As discussed at the beginning of the chapter, everyday examples include waterfalls, whirlpools, clouds, and lightning. Given their commonplace and ongoing existence, some might question the importance of the theory of dissipative structures. In this context, however, its discovery is akin to the discovery of the Americas.

The Americas were not new for their indigenous inhabitants, of course, who had lived there for tens of thousands of years. The importance of the discovery of the New World was that it was new to Europeans. The discovery enabled modern European civilization to greatly extend its territory, and the new seeds of European culture planted in the New World grew so fast, successfully, and powerfully that they completely merged with and enriched Western culture. Old World culture expanded beyond its historic European, West Asian, and North African roots.

Comparing Prigogine to Columbus demonstrates the importance that the discovery of dissipative structures has to science, medicine, and modern civilization. While dissipative structures existed before Prigogine and modern scientific theory, their discovery by science greatly expanded its territory and that of the disciplines of physics, chemistry, biology, and medicine.

In addition to enabling the mysteries of acupuncture and other traditional forms of medicine, which had been discovered intuitively in ancient times, to be rationally understood in the eyes of modern science, this rediscovery of old "new structures" will also facilitate a paradigm shift in scientific thinking. It will enable venturing beyond the old continent of conventional Western medicine, with its associated reductionist thinking, to the simultaneously ancient and new continent of holistic medicine, with its new worldview and inclusive thinking. Furthermore, this new way of thinking and new outlook will profoundly influence not only the development of medical technology, but also the direction of the development of our civilization.

CHAPTER 8

STANDING WAVES AND WAVE SUPERPOSITION

With the music of stringed instruments and with melody on the harp, because of what you have done, I sing for joy. —PSALMS 92:3–4

It is ironic that Columbus never knew that he had discovered a new continent; he firmly believed what he found was a new route to India. The task of naming the new continent and exploring the land was done incrementally by many others over the subsequent five hundred years. Prigogine was arguably much luckier in knowing he had discovered a new realm and witnessing the explorations of others in the new territory.

Standing Waves Are Dissipative Structures

The conventional standing wave, described in chapter 3 and known in physics for hundreds of years, is also a kind of dissipative structure. Understanding how these waves work and some of their properties is key to comprehending how acupuncture can influence the dissipative structure in the body.

The first image in figure 8.1 depicts a simple example of a standing wave on a string. It resembles three spindles placed head to tail. These are not solid structures, like real spindles, but dynamic structures. The more precisely a standing wave is observed, the more apparent its instability becomes. This is especially true for relatively high frequencies.

The energy supporting the dynamic structure is provided by a small motor that keeps the string in permanent vibration, as shown in the second image. The standing wave also constantly dissipates energy, mostly in the form of acoustic waves. The length of the string determines the frequency of the standing wave,

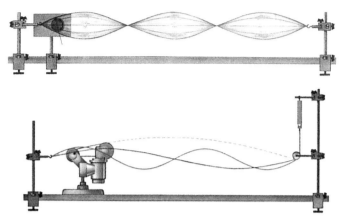

Figure 8.1. Standing wave in a string (top) and the energy supplied by a motor (bottom).

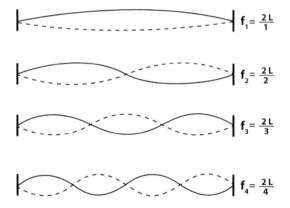

Figure 8.2. The fundamental frequency (f_1) and the frequencies of the overtones (f_2, f_3, f_4) are determined by the length of the string.

shown in figure 8.2, which in turn determines the frequencies of the related acoustic waves emitted by the string. All stringed musical instruments function according to this working principle.

In musical instruments, the length of the string is controlled by the fingers of the musician. Constantly changing finger positions allows the musician to establish standing waves of desired frequencies at specific times, thereby producing the melody. In most cases, several different frequencies exist together on the same string, as shown in figure 8.2. The lowest frequency, f_1, is called the fundamental frequency and is usually the one written in the score, indicated by the note. In musical terms, the frequencies f_2, f_3, and f_4 are referred to as overtones. Different musical instruments produce varying combinations of strengths of individual overtones. This creates what is known as the timbre or tone color of an instrument, and distinguishes the sound of one violin from another, or a violin from a guitar.

The working principle of wind instruments is also based on the regulation of standing waves. The tube of a wind instrument provides a resonance cavity in which there are invisible standing waves of varying air pressure. The frequencies of these standing waves are determined by the length of the tube. By altering finger placement to alternatively open and close holes in the tube, the musician changes the effective length of the tube. This modifies the frequency of the standing waves inside the tube, thereby changing the note emitted by the instrument.

The frequencies of standing waves in percussion instruments are dictated by the size of the instrument. As it is not easy to change the size of the resonance cavity in a percussion instrument, performers usually use a series of percussion instruments that correspond to various notes to play music.

Superposition of Standing Waves

As mentioned earlier, waves possess another interesting feature: unlike particles, two or more waves can simultaneously occupy the same space, combine together to form a single new wave, or even cancel out one another completely, vanishing into nothing. In other words, they can superimpose. In the terminology of physics, they interfere with one another, thereby forming a new interference wave. As such, physicists refer to the phenomenon as interference.

Superposition of Two Waves

First, let us consider what happens when two waves, with the same wavelength and the same phase—their peaks and troughs occurring at the same time and place—occupy the same place. Figure 8.3 demonstrates the outcome; a new wave is created with a larger amplitude (height). The amplitude of the new wave is equal to the sum of the two original amplitudes.

It is not difficult to visualize and understand this addition operation in the procedure of superposition. It is, however, somewhat unusual that the arithmetic in the world of waves is 1 + 1 = 1, at odds with the ordinary arithmetic of 1 +

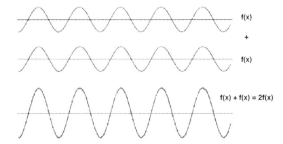

Figure 8.3. Superposition of two waves in the same phase: constructive interference.

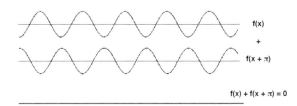

Figure 8.4. Superposition of two waves with phase difference of half a cycle: destructive interference.

1 = 2 in the world of particles. In a way, it is like two people sitting on a couch together becoming a new person of double size.

When two waves with opposite phases are superimposed in the language of mathematics, the two waves have 1π (180° or half a cycle) phase difference, or to put it more simply, when one wave is at a peak, the other wave is at a trough. The outcome of the superposition is known as destructive interference, and the new wave has zero amplitude. The peaks of the first wave coincide with the valleys of the other wave, cancelling each other out. This is the working mechanism behind noise-canceling headphones and other forms of active noise reduction. In this case, the arithmetic is even more unusual: 1 + 1 = 0. It seems that the two people sitting together on a couch disappear.

In terms of physics, the first case (fig. 8.3) is called constructive interference because the outcome is greater than the two original waves. The second scenario (fig. 8.4) is called destructive interference because the outcome is smaller than the two original waves. These examples are the extremes; in reality, most cases of superposition and interference occur somewhere between them. Examples include figure 3.5 in chapter 3, which shows the classic double-slit experiment—two light beams may enhance each other, creating

Standing Waves and Wave Superposition

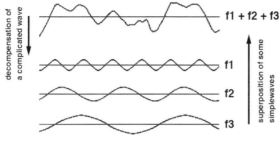

Figure 8.5.
A complicated curve from the superposition of three waves.

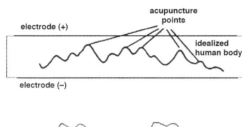

Figure 8.6.
Comparison of the curve from electronic measurement on the human body and the curve created by superposition of three waves.

brighter regions (constructive interference), or cancel each other, producing darker regions (destructive interference). Figure 3.4 shows what happens when two waves of different wavelengths interact—a new frequency, called a beat frequency, is generated by the superposition.

Superposition of Many Waves

The process of superposition is not limited to two waves; any number of waves can interact, as in figure 8.5. By superimposing a number of simple waves, it is possible to create any number of periodic curves, which are a continuous repetition of the same shape, as in the beat frequency in figure 3.4, or even aperiodic waves, at the top in fig. 8.5.

In light of this understanding, it is useful to revisit the background of electronic measurements on the skin. As explained in chapter 6, these measurements actually reflect the distribution of energy inside the body. Figure 8.6 allows for comparison between the curve in electronic measurements on the human body (top in fig. 8.6), and the curve generated by the superposition of three waves (bottom in fig. 8.6). This makes it easier to visualize how a heterogeneous distribution of the electromagnetic field could be generated by the

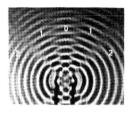 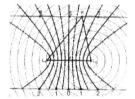

Figure 8.7. Interference pattern and calculations for two two-dimensional circular waves.

superposition of electromagnetic waves. In reality, the complexity of the human body and the multitude of electromagnetic wave emitters in it result in much more complexity than the simplified conditions in figure 8.5.

Two-Dimensional and Three-Dimensional Standing Wave Interference

The preceding discussion only considered superposition and interference of waves and standing waves in one dimension. In reality, most situations involve waves in two or three dimensions. In theory it is possible to calculate interference patterns in two- and three-dimensional cases (fig. 8.7). Practically, however, even with powerful computers, these calculations are usually a mammoth undertaking. Fortunately, the work required to observe interference patterns in two dimensions is not overly complicated. In chapter 3, figure 3.7 depicts interference patterns of standing waves in a vibrating disk, and in figure 3.8, inside a violin.

Modulation of the Interference Pattern

Once it is established that the subtle energy distribution inside the body is mostly determined by the superposition of electromagnetic waves, that is, the energy distribution is in fact an interference pattern, a method of modulating it becomes apparent. An unhealthy energy distribution inside the body of a patient can be improved by intentionally introducing a disturbance into the interference pattern. This can alter the energy distribution, thereby allowing the patient's health to improve.

Changing Frequencies

Medical doctors in Western countries have already invented numerous methods of introducing electric stimulation to a patient's body to assist in regaining health.

While these doctors were unaware that this stimulation would alter the patient's energy distribution, they were comfortable with the term *energy medicine*. They believed they had detected some pathogenic bioelectricity in patients. As such, they thought they could cancel it out by introducing external electrical stimulation.

These pioneers ventured beyond the chemical aspect of the body into the energetic aspect of medicine, an important step forward that contributed greatly to the development of medicine. Unfortunately, these methods and their inherent way of thinking are still limited by the framework of Western medicine and reductionism. If the effect of electrical stimulation on the existing interference pattern of electromagnetic standing waves in the body were better known, it would be understood that the treatment is holistic. Moving toward a better understanding of the source of the heterogeneous energy distribution in the body will inevitably lead to significant developments in energy medicine.

Changing the Boundary Condition of a Resonance Cavity

In addition to introducing external electrical or magnetic stimulation, the interference pattern of electromagnetic waves in the body—the energy distribution—can also be modulated by changing the boundary conditions of the resonance cavity. As is evident in figure 3.8 in chapter 3, the shape of the interference pattern of standing waves produced inside a violin is not solely determined by the frequency of stimulation, but also by the shape of the cavity.

To further illustrate, consider interference patterns of standing waves on the surface of water (plate 11 in color plate section) created by placing a loudspeaker beneath a pan of water. When the loudspeaker emits a constant sound signal, a stable interference pattern appears. At left, a ring appears in the water. At right, three smaller rings of varying diameter have been introduced instead. In both cases, the frequency of the sound signal was the same. The interference pattern was greatly modified by means of simple change in the boundary conditions.

Significant changes in interference patterns do not require major changes in boundary conditions. Sometimes a tiny change can greatly alter the interference pattern and related frequencies that form standing waves. For example, it is well established that a small slit will greatly damage the sound of a musical instrument.

Physicist Hans-Jürgen Stöckmann, from Marburg University in Germany, conducted both experimental research and theoretical calculations concerning the change of standing waves on a rectangular shape. His results show that

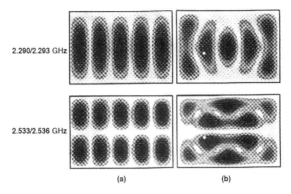

Figure 8.9. Interference pattern in a rectangular plate (a) without a hole and (b) with a hole (the white dot).

even a tiny hole in the shape greatly changes the interference patterns that arise (fig. 8.9).[1] In order to get the most pronounced changes in the interference pattern, specific points need to be chosen to make the holes. The best places are in regions where the energy of the original interference pattern is highest. This provides insight into why acupuncturists always insert their needles at locations with lowest skin resistance and call these sites acu-points. The locations exhibiting the lowest readings for skin resistance are actually the places of highest energy, determined by the interference pattern of the standing electromagnetic waves within the body, which is a dissipative structure of the electromagnetic field inside the body.

Figure 8.10 is a simplified one-dimensional example to illustrate how an acupuncture needle may work. Suppose the electromagnetic waves being emitted by an ailing organ have altered from their original frequency to an inappropriate one. In turn, this wave has generated a detrimental standing wave. Inserting a needle at the acu-point establishes a new boundary condition at one of the peaks of the electromagnetic standing wave. As the best place to insert a needle to cancel a standing wave is at any peak, this will serve to cancel the inappropriate standing wave. As a result, the unwell organ will be positively influenced, helping it return to its optimal frequency.

Of course, real conditions are far more complicated than the idealized situation in figure 8.10. There are multitudes of standing waves in the body that form a complex interrelated three-dimensional interference pattern. Consequently,

Standing Waves and Wave Superposition

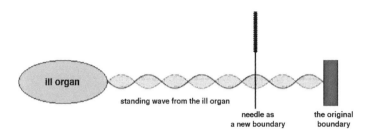

Figure 8.10. A possible working mechanism for the effect of needling.

inserting a needle into a point of high-energy intensity in the body introduces a major disturbance to the entire interference pattern.

The change in interference pattern during acupuncture, or any other disturbance, can be monitored by attaching a probe matrix to the skin. The results can then be intelligibly expressed using pseudo-colors on a computer display (plates 11 and 12 in the color plate section). The images show how the interference pattern in the arm changes over the time that a needle remains in an acu-point. This experiment also demonstrates that if the needling operation is successful, it generally takes between ten and twenty minutes for an interference pattern to change from its initial stable state to a new stable state. This period is consistent with the duration of acupuncture treatment sessions.

The slow pace of the preceding change in the interference pattern is of special importance. As discussed in chapter 5, physicist Bi-wu Zhang pointed out in 1959, long before the discovery of dissipative structures, that the notably slow speed of sensation propagation along meridians, in contrast to the extremely high speed of individual electromagnetic waves, could come from the grouping speed of many electromagnetic waves. The situation depicted in figure 8.11 involves the interactions of billions and billions of electromagnetic waves. As such, the speed of signal progression becomes so slow, even for the small area being monitored, that it takes twenty minutes for the change in the interference pattern to complete.

CHAPTER 9

WIRELESS COMMUNICATION INSIDE A BODY

When we magnify our physical matter very much, we find that we are mostly of void permeated by oscillating fields. This is what objective physical reality is composed of. —ITZHAK BENTOV

Living systems like our bodies are complicated and require well-developed communication systems to maintain homeostasis, to keep coherence between organs, and to produce appropriate reactions to external disturbances and to changeable surroundings. The many kinds of communication systems inside a living organism can be divided into three categories.

Postal Communication: Chemical Substances

The first category of communication within living systems is similar to postal communication. Perhaps civilization's oldest communication system, it requires substantial carriers of information, such as letters, postcards, and parcels, as well as a clear address for the recipient. Our bodies employ chemical compounds to facilitate internal communication in an essentially similar manner.

Chemical communication inside living organisms can be described in the context of a key-and-lock model (fig. 9.1). Information is issued by a gland, the sender, and is carried by a hormone, the messenger, via blood or tissue fluid to a target cell, the recipient. In this system, the address of the recipient is encoded on the messenger by the sender, ensuring that the messenger can only be received by a specific recipient. This is achieved by encoding the address in the form of a chemical pattern so that it interacts with the receptor in a manner similar to a key in a lock.

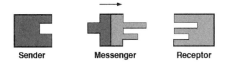

Figure 9.1. Postal-style communication in the body: the key-and-lock model of chemical communication.

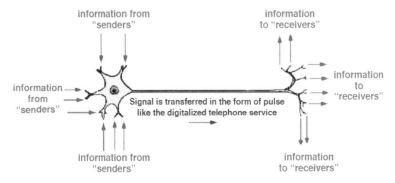

Figure 9.2. Telephone-style communication in the body: signals in cables.

This mechanism is employed by the endocrine system, the immune system, and many medications. Being relatively mechanical in nature, this model is conceptually straightforward and understandable. It serves as the classical model that currently dominates the thinking of people in general, medical doctors, and, of course, the pharmaceutical industry.

Telegraph and Telephones: Signals in Nerve Fibers

The invention of the telegraph by Samuel Morse (1791–1871) and more significantly the invention of the telephone by Alexander Graham Bell (1847–1922) introduced humanity to a second communication system. People's voices and other information were transported directly over long distances by wires. Scientists found a system corresponding to telegraph or telephone communication in our bodies: the nervous system (fig. 9.2).

Wireless Ethereal Communication

Marconi's invention of the wireless long-distance telegraph had profound implications, opening the possibility of communicating through ethereal fields and waves. As discussed in chapter 3, ancient societies believed in the existence of ghosts more than contemporary people do, but today we believe in the existence of electromagnetic fields, even though they are invisible, and the existence of electromagnetic waves, even though they are inaudible.

There is no essential difference between ghosts and electromagnetic fields, except that fields can be quantified by mathematical formulas, the Maxwell equations, which quantitatively describe their actions. This enables us to manipulate electromagnetic fields by means of electronics. As yet, we have not found mathematical formulas to describe the behavior of spirit, so no standardized technique currently exists to manipulate it.

In essence, electromagnetic fields possess ethereal qualities. Perhaps people in ancient times were somehow sensing electromagnetic fields and waves that existed outside the range of everyday perception. Not having access to science and mathematics to describe them or the technology to manipulate them, their worldview might attribute the experience to ghosts.

The realization that the electromagnetic waves that facilitate our communication in a sense represent a domesticated ethereal realm can be somewhat disconcerting. Even when our TVs, radios, and mobile phones are switched off, our space is haunted and completely occupied by these entities. Invisible and silent, these apparitions permanently inhabit our rooms, beds, and desks. They even occupy the space between people in direct conversation. While we never see or hear them, they are constantly moving and making noise.

Ancient beliefs held that ghosts sometimes kept silent and sometimes spoke, sometimes were invisible and sometimes became visible. Today, modern technology makes these electromagnetic entities audible and visible.

Wireless Communication in Living Systems

Scientific research into wireless communication in living systems lags far behind the development of wireless communication technology. While lots of children use mobile phones and wireless internet surrounds us most of the time, scientists in biology and medicine are still fixated on the body's postal communication—through molecules—and on telephone communication—through nerve fibers.

There are, of course, technical challenges in this kind of research, but the biggest problem is conceptual. For instance, in chemical communication, the interaction between an actual messenger, such as a hormone molecule, and a

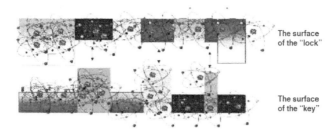

Figure 9.3. Selective electromagnetic interaction between "key" and "lock" molecules.

receptor, the surface of a target cell, is not mechanical at all but rather is electromagnetic (fig. 9.3). The key-and-lock model gives the misconception that the interaction between the messenger molecule and the receptor molecule occurs through mechanical force. At such a microscopic scale, mechanical force does not play a significant role; the interface occurs through electromagnetic interaction. The dominance of the mechanical key-and-lock model impeded our understanding of the wireless communication occurring between messenger and receptor. As such, it can be argued that it has misled the direction of medical research and obstructed development in medicine.

Research into wireless communication between living organisms actually started very early. In 1922, Russian biologist Alexander Gurwitsch (1874–1954) performed the renowned "onion experiment" (plate 13 in the color plate section). Two onions each have all but one of their roots removed. The remaining roots are then placed in glass tubes. The glass tube holding onion 2's root, which acts as the receiver, is sheathed inside a metal tube in order to shield the root from disturbance by outside electromagnetic waves. As shown at right in figure 9.4, there is a hole in the metal and glass tubes where the tip of the root of onion 1 is located. Gurwitsch found a greater rate of cell division on the side of onion 2's root facing the hole than on the side facing away. This showed that there was some signal triggering cell division being transferred from the tip of the root of onion 1 to the skin of the root of onion 2. Upon further investigation, Gurwitsch found that this effect did not occur when ordinary glass was placed between the two roots but did with quartz glass between them. Quartz glass does not block ultraviolet rays, while normal glass does, leading him to postulate that the communication occurring between roots might be in the ultraviolet range.

This experiment served as the first step in proving the existence of wireless communication and providing the approximate frequency range of electromagnetic waves that carry the signal. Plenty of work remained to be done in determining the precise band of electromagnetic frequencies utilized and how signals were encoded among and inside living systems.

Unfortunately, this avenue of research was interrupted by the two world wars. After World War II, development in biology directed at exploring particles and molecular biology flourished. Biologists of the last half century have focused their attention on the postal and wired communication systems in living entities and completely overlooked the existence of wireless communication.

As discussed in chapter 7, the development of molecular biology is already past its zenith. A huge amount of parallel supplementary work is still required, but there remains little chance of major breakthroughs, like the discovery of the double helix structure of DNA half a century ago, in this area. There is also a growing awareness among medical doctors of the limitations and weaknesses of contemporary molecular medicine. Consequently, more attention is being paid to the energetic aspect of medicine, including therapies such as soft-laser stimulation, colored-light stimulation, extra-weak microwave stimulation, and electric pulse puncture. These work by introducing various electromagnetic signals into the body to regulate and harmonize patients' dissipative structure.

Similarly, there is also a growing awareness of the limitations of contemporary molecular biology among biologists. This is reflected in increased use of terms such as *morphogenetic field* and *biofield*. The concept of a morphogenetic field was suggested after biologists discovered that the process of embryonic development is not solely dependent on genes but is also guided or controlled by invisible factors. The existence of these factors is particularly apparent in the wound-healing process. For instance, in salamander tail regeneration, it is easy to see that an invisible fine blueprint had predetermined not only the shape of the salamander tail but also its fine structure. Biologists called this invisible fine blueprint the morphogenetic field or biofield.

Increased consideration of fields has seen more scientists, particularly physicists, venture into biology. In some ways this is reminiscent of the 1960s, when chemists became involved in biology and greatly advanced biochemistry and molecular biology. Generally, physicists are averse to introducing the concept of new fields and will not do so unless it is unavoidable. From the viewpoint of physicists, the number of factors should be reduced to a minimum. What is being referred to as the morphogenetic field or biofield is likely to be a mixture of many different interactions, including instances of chemical interaction, mechanical interaction, temperature interaction, and electromagnetic interaction. Serious study of the field aspect of living systems should study each of these interactions individually. Among the four categories, electromagnetic interactions play the most important role in the energetic aspect of living systems.

Understanding the chemical aspect of living systems is undeniably important and arguably deserved to be the first aspect to be thoroughly investigated. Now that this investigation is nearing completion, we should consider how to progress to the next step. This involves venturing into a more complicated and possibly more important aspect, that of the electromagnetic interaction in living systems.

Electromagnetic Body versus Chemical Body

To clarify the present situation and to visualize the next step, an artificial and conceptual separation between the chemical body and the electromagnetic body will be made.

The Chemical Body

We are already well acquainted with the chemical body, which consists of solid bones, muscles, and organs linked together by blood vessels and nerve fibers. All of these are composed of cells, which are in turn made of proteins, DNA, RNA, enzymes, coenzymes, and countless small molecules and ions. The chemical body is a solid pattern of particles and has been elaborately studied during the marvelous development of molecular biology in the last half century. It seems that no physical aspect of our bodies is unknown to us. As such, modern medicine based on the study and understanding of the solid chemical body can be considered classical.

The Electromagnetic Body

In addition to the chemical body, there exists an electromagnetic body, and in my opinion, it plays a role of equal or possibly greater importance. It represents an unknown territory in our bodies that now lies within the reach of modern science. Having moved from the realms of mystical experience, science fiction, or pure speculation, it is becoming an important area of basic scientific research in biology and medicine.

The electromagnetic body is much more complicated and dynamic than the chemical one. If we were able to see it, it would appear completely different to the visible chemical body. We would observe the seven major chakras along the central line of the body and many small chakras in other places emanating various colors. We would see dozens of acupuncture meridians, hundreds of acupuncture points, and numerous micro-acupuncture meridians and points weaving into an intricate network—a continuous interference pattern that is holographic in appearance. Around the body we would discern the aura, as described in ancient beliefs. It exists to some extent in the detectable range of extra-weak light as well as in the infrared and microwave parts of the electromagnetic spectrum. Modern technology has seen this become a new area of serious scientific research.

In addition to being highly complex, the electromagnetic body is also extremely dynamic. Unlike the chemical body, where the bones, organs, vessels, and fibers have fixed positions, definite volumes, and distinct boundaries, the electromagnetic "organs," such as chakras, and some invisible "vessels," such as acupuncture meridians, exhibit only a relatively stable position with nebulous boundaries and variable volumes. They are continuously flashing and exhibit continuous changes in intensity, color, and shape, like the surface of the ocean in a fierce storm. This turbulence is particularly evident when the person is experiencing an intense change of emotions and psychological states.

If we were able to perceive the electromagnetic field in greater detail, we would witness tremendously complicated communication processes being performed at extremely fast speeds. Electromagnetic waves and photons facilitate communication inside the cells, between cells, between bodies, and with the surroundings. All this occurs in addition to the communication occurring through nerve fibers, hormones, and other molecules. As with wireless communication and television broadcasting, communication through electromagnetic fields carries much more information over much wider channels than can be transmitted through insulated nerve fibers and the slow interaction between the surfaces of molecules. Thus, communication within the electromagnetic body has a more profound and subtle influence on our bodies and health.

Unique Challenges of Studying the Electromagnetic Body

The scope of our senses inevitably dictated the course that science and medicine took in investigating our bodies. Starting with the visible and material aspects, it later began to slowly venture into the invisible and field parts. This latter exploration poses unique challenges that exploring the solid physical realm did not.

Firstly, like the radio waves that continuously enshroud us, the electromagnetic body is invisible. Apart from the narrow range of electromagnetic waves that we can perceive as visible light, we can only visualize their structure and pattern in our imagination. We are unable to study the electromagnetic body using our eyes, microscopes, or chemical analyses. For this reason, even the existence of an electromagnetic body was, until recently, an open question in biology and medicine. Compared to the accessibility of the chemical body to direct perception, the challenge of inferring the existence of the electromagnetic body through meticulous analyses and synthesis of complex experimental data is daunting.

The current situation in biology and medicine is similar to that of physics in the nineteenth century, when people were confronted with evidence of the existence of invisible radio waves. People did not believe in their existence even though the world was full of them. It took Faraday discovering the relationship between electricity and magnetism, Maxwell discerning the formula to describe the relationship and thereby predicting the existence of electromagnetic waves, and finally Marconi inventing long-distance radio transmission to convince people that they existed. While their existence is established in today's world, we are still unable to perceive them directly. The invisibility of the electromagnetic body poses an enduring challenge to its research and recognition.

The second problem is that the electromagnetic body is highly dynamic, sometimes like an ocean in a storm, making it harder to observe, measure, and formulate laws of its behavior. Also, the more precise the measurements become, the more unstable the resulting values are. In addition to continuously fluctuating with varying intensity and frequency, the electromagnetic body also changes shape in response to a change of location, and it is modified especially by various pathological, physiological, and psychological states. The images and the concepts of the electromagnetic body are fundamentally different from those of the material chemical body. Its dynamic, variable nature, at odds with the ingrained notion that we have a stable solid body, make it more difficult to recognize. In addition, formulating a theoretical system to mathematically describe the characteristics and movement of such a highly dynamic structure poses a great challenge.

Thirdly, we are unable to study the electromagnetic body using the conventional method of separation and isolation used in anatomy, or by using an electron microscope or chromatography, which allows an organic mixture to be separated into its thousands of different constituent compounds. The electromagnetic body is both inseparable and untouchable. Due to the strong effect that a conductive metal has on electromagnetic fields, even bringing a dissecting scalpel close to the body can greatly disturb the refined structure of the electromagnetic body. If the scalpel cuts into the body, it can cause even greater damage to the original structure of the electromagnetic body. The more a body or a cell is divided into pieces, the more damage is done to the electromagnetic body. If the process of division were carried through to the molecular level, the structure of the electromagnetic body would be completely destroyed. Consequently, the conventional approach of reductionism, which has been so successful in the study of the chemical body and molecular biology, does not work at all in the study of the electromagnetic body. It is not only impossible to use a knife to cut into the body but also impossible to insert any detecting probe into the body without causing major disturbance in the subtle electromagnetic body.

The final challenge is that electromagnetic waves travel more than a million times faster than the movement of molecules and nervous system pulses. Consequently, the ergodic phenomenon occurs—where, given a sufficiently long time, a system can travel through all possible states. In the case of the electromagnetic body, it arises because electromagnetic waves travel so fast that they can cross the body numerous times and share all information through all parts of the body. For this reason, no event in the electromagnetic body can be isolated from the rest of the body, and the information from every event can be found in the whole body—it is completely holistic.

It is impossible to study the electromagnetic body using the analytical methods that have been so successful and important in the development of science. Our habitual modes of thinking make it hard to enter a new field but also to move away from the conventional mode of separation, isolation, and analysis. That being said, moving on is required if we are to search for a completely new framework and its related methods to study this new area of science and medicine.

Practical Solutions for These Challenges

Technological development plays a key role in furthering our understanding of the electromagnetic body. Since the electromagnetic field in the body is very weak and prone to disturbance, a sensitive detection system is required to discern its subtle structure. In order to reduce unnecessary disturbance, a remote detection system is preferable. This passive detection system would not introduce anything to the body, such as electrical current in the case of resistance measurements. Fortunately, the last five decades have seen the development of many sensitive detection technologies—for example, technologies originally developed for satellites have been applied to solving medical and biological problems. This implies that the technological basis required for a serious study of the electromagnetic body is already in place.

On the other hand, new theories and methodologies for data analysis, or more precisely, for data synthesis, are required to cope with the extremely dynamic and unstable measurements of the ceaselessly fluctuating electromagnetic body. Holistically investigating a complex, highly interrelated dynamic system is inherently more challenging than the relatively straightforward approach of reducing a stable structure into finer and finer components. Fortunately, in the last two decades, the study of nonlinear problems has offered many new methods, such as bifurcation theory, catastrophe theory, chaos theory, fractals, coherence theory, and new theories of statistics and other mathematics. Meanwhile, the continued rapid development of computer technology offers a powerful tool to deal with a quantity of data that previously would have been unmanageable.

On a side note, perhaps the most important thing to bear in mind is that, in my opinion, biology and medicine are currently in a similar stage as physics was a hundred years ago when it progressed from classical physics to electrodynamics and quantum physics. It is disconcerting to change from the familiar research into solid objects, with their visible and reliable properties, to an uncharted area of intangible objects that are inconstant and inferred from data. We have become familiar with the initially disturbing concepts of subatomic particles and black holes over time but are not yet used to imagining the intangible part of our bodies. To address this problem, it is necessary to reinforce biology and medical education in mathematics and modern physics in order to complement their already strong education in chemistry and biochemistry. Also, more physicists and mathematicians should be involved in biological and medical research teams. This will help introduce the theoretical basis for the study of the electromagnetic body, which is established in mathematics and modern physics, into the study of biology and medicine.

The Electromagnetic Body as a Common Foundation for Complementary Medicines

In light of the electromagnetic body, many puzzling problems and mechanisms in the complementary medicines become understandable. These include the nature of acupuncture meridians and points, the mechanism of homeopathy and other holistic therapies, and even the mysterious experiences that feature in ancient medical traditions.

The many failures in the search for acupuncture meridians and points by means of anatomy in the last half century have already been discussed in detail. The reason behind these failures is that the structure of acupuncture does not exist in the chemical body. However, in light of the electromagnetic body, the acupuncture system can be understood as the prominent areas of an interference pattern formed by superposition of invisible electromagnetic standing waves. In this context, many puzzling problems can be understood, including the shape, size, and stability of acupuncture points and meridians, the relationship between organs and acu-points, the relationship between anatomic structure and distribution of meridians and acu-points, the effect of needling without remedy, the transmission of a signal along meridians and its speed, the phenomena of bio-holography, and the statistical self-similarity of conductivity measurements.

Homeopathy, which is based on the principle of similarity and the potency rule, has also been a great mystery in medicine. The principle of similarity means a substance that causes the symptoms of a disease in healthy people will cure similar symptoms in sick people. The potency rule means that remedies are prepared by repeatedly diluting a chosen substance in alcohol or distilled water, followed by forcefully striking it on an elastic body, called "succussion." Each dilution followed by succussion is said to increase the remedy's potency.

British physicist Cyril W. Smith spent much of his life engaged in serious and systematic research into the mechanism of homeopathy. He asserted that the solution for the mechanism could only be found by considering the electromagnetic structure within the water. If we consider the electromagnetic body, the mechanism of homeopathy can be understood in terms of the resonance effect in the electromagnetic body. As discussed, there are numerous electromagnetic oscillators in a human body. These act as the sources of the numerous electromagnetic waves that construct the extremely complicated, dynamic structure of the electromagnetic body through infinite reflection and superposition. There are many interactions between the electromagnetic structure in the water of a homeopathy remedy and the structure of the electromagnetic body by means of weakly coupled oscillators through the electromagnetic field. The coupling and the energy transfer between them fits into the principle of similarity and the potency rule of homeopathy.

Given this workable explanation, the only challenge remaining is to find an extremely sensitive instrument capable of detecting such a weak resonance effect. The human body itself is the most sensitive detector for the extra-weak signal from a homeopathy remedy. Smith demonstrated this with very good reproducibility through subjective methods like dowsing and swinging a pendulum, but it is difficult for the scientific community to accept techniques as subjective as these. The human body is also a good amplifier of electromagnetic interaction, and the structure or pattern of the electromagnetic body is very sensitive to any electromagnetic disturbance. Therefore, it is possible to objectively detect the significant change in the interference pattern in the electromagnetic body that is caused by an extra-weak disturbance.

The concept of chakras, which means "light rings," is central to traditional Indian medicine, which holds that the condition of the chakras is related to the psychological and physiological state of the body. From the viewpoint of the electromagnetic body, the chakras are the focal points of various waves, in particular the electromagnetic waves in the body. Therefore, the chakras belong to the interference pattern of electromagnetic waves, which is strongly connected with the psychological and physiological state of the body.

In Traditional Chinese Medicine, emotion is regarded as the primary cause of the majority of diseases. In contrast, classical Western medicine considers the human body to be a complicated machine, and the medical doctor a sophisticated mechanic. This view leaves little space for emotion. However, the last two decades have seen psychosomatic medicine becoming more important while psychology now plays a more important role in health care and medical insurance. For a long time, psychologists have taken great efforts to find explanations of psychological states in terms of physiology and other branches of biology without clear success. The problem is that the dense chemical body has only a very limited and indirect connection to psychological states. The state of the electromagnetic body, in particular communication through the electromagnetic field, has a closer connection to psychological and emotional states and is much more sensitive to emotion and environmental influences, changing long before any substantial pathological change would occur in the chemical body. Effective evaluation of the state of the electromagnetic body promises many advantages for the health care and medical insurance industries, because problems appear in the more dynamic electromagnetic body before any substantial changes occur in the chemical body. Consequently some conditions could be diagnosed and rectified earlier, thereby alleviating the need for hospitalization later on.

In my opinion, we are currently at the dawn of a new era that requires progressing from the old world of research, with its focus on the chemical body and allopathic medicine. It has a formidable legacy earned through saving countless lives through surgery, antibiotics, and other drugs, but to improve quality of life and counter so-called modern diseases, we have to venture into a new world that considers the electromagnetic body and regulation medicine. Greater awareness and understanding of the electromagnetic body will facilitate more effective use of self-regulation methods such as physical exercise and meditation.

The distinction being made here between the electromagnetic body and the chemical body is to facilitate a simple and clear depiction of the situation. There is in fact no separation between the two—they are interconnected and interact perpetually.

Indian philosophy holds that people are composed of seven levels, with each level progressively less dense and harder to see. From this viewpoint, the chemical body may only represent the first level, and the electromagnetic body the second level. At this point there is no clear scientific evidence to definitively state whether or not there are any other bodies beyond the chemical body and the electromagnetic body. However, in comparison to the long history of humanity, science, at three hundred years old, is only in its infancy. It has graduated from theorizing a world comprised of dense matter to one including fields; the next step is for our generation to progress from a biological understanding based on the chemical body and pharmacological medicine to that of the electromagnetic body and regulation medicine. Will there be progress beyond this? It seems certain.

PART 4
FIELD AND WAVE ASPECTS OF BIOLOGY

CHAPTER 10

POWERFUL RESONANCE: A SECRET MEANS OF TRANSFERRING ENERGY AND INFORMATION

The great extension of our experience in recent years has brought to light the insufficiency of our simple mechanical conceptions and, as a consequence, has shaken the foundation on which the customary interpretation of observation was based. —NIELS BOHR

The Secret of Resonance

Resonance is not a new concept in physics, but it has only recently begun to be considered in biology and medicine. Within biology's focus on the composite of particles that form the chemical aspect of biology, resonance plays almost no role at all. Biologists and medical doctors are relatively unfamiliar with resonance and its charming characteristics.

When the focus is shifted to waves, in particular to the field aspect of biology, it becomes apparent that resonance plays a key role in many living processes. In fact, many mysterious phenomena in living systems can only be understood in light of the characteristics of resonance. Further development of our understanding of the wave aspect and the resonance effect in living processes would enable a completely new appreciation of how intriguing living processes can be.

As resonance is an established concept in acoustics and in music, this field can provide us with some fundamental insights into the peculiar characteristics of resonance, to illustrate the role that electromagnetic waves can play in living systems.

Figure 10.1. A child applies the resonance effect in order to swing high with little effort.

Accumulating Energy

The first characteristic of resonance to consider is that it enables energy to slowly and imperceptibly accumulate by increments. The underlying mechanism of resonance can be understood by considering how a child plays on a swing. The first push or pull does not move the swing much. However, large movements can ultimately be achieved with the repeated application of a small force.

The secret to successfully accumulating energy from the repeated application of small amounts of force is that the frequency with which they are exerted should be the same as the natural frequency of the swing—the frequency it will oscillate at if there are no forces driving it or slowing it down—and in the same direction as its movement. Children intuitively learn how to harness this resonance through experience.

The accumulation of energy brought about by resonance sometimes poses a danger. Soldiers are required to "break step" when crossing a bridge to avoid the possibility of their marching frequency matching the bridge's natural frequency, thereby damaging or destroying the bridge (fig. 10.2). Even when precautions are taken, bridge failure can occur. In 1850 the 335-foot-long Angers Bridge in France collapsed while a battalion of soldiers was marching across, killing 226 of them. They had arrived during a powerful storm, and the wind had caused the bridge to sway. In an effort to keep their balance on the swaying bridge, they may have inadvertently stepped with the same frequency that the bridge was swaying at, resulting in the collapse.

A more recent example of this phenomenon, sometimes called a resonance catastrophe, was captured on film in 1940 when the Tacoma Narrows Bridge in

Powerful Resonance: A Secret Means of Transferring Energy and Information

Figure 10.2. Resonance catastrophe: the consequence of the accumulation of small amounts of energy.

Washington State failed. In this case, the wind caused the bridge to oscillate at one of its natural frequencies, leading to its collapse.

Resonance catastrophe also poses a serious danger in airplanes and rockets. If a source of mechanical vibration in an airplane or rocket is the same as the vehicle's natural frequency, serious and even catastrophic damage can occur. Two crashes involving the Lockheed L-188 Electra, in September 1959 and March 1960, were caused by weak engine mounts producing a vibration that matched the natural frequency of the wings, causing them to tear off in mid-flight.[1] These crashes serve to illustrate how powerful the accumulation of energy through resonance can be. While the energy of the individual waves in the human body is very small, the possibility of danger inside the human body remains. What would be the effect of a resonance catastrophe occurring inside the mind-body system?

Transferring Energy

The combination of waves and the resonance effect allows energy to be subtly transferred from one body to another. This can occur over very long distances, even millions of miles, and through a vacuum. We can understand this characteristic of resonance by considering the acoustic interaction between two tuning forks (fig. 10.3). The precondition for resonance between the two is that they must have precisely the same natural frequency.

Suppose that tuning fork 1 is vibrating with a large amount of energy. Its vibrating energy will gradually dissipate as it pushes and pulls the air molecules

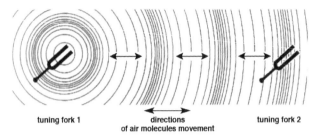

Figure 10.3. Energy incrementally transferred between tuning forks.

surrounding it to produce acoustic waves. Tuning fork 2, which was initially silent, receives a sound signal from tuning fork 1. The pressure variations that accompany the movement of molecules in the sound wave will cause the fork to be pushed and pulled with the same frequency as tuning fork 1. As tuning fork 2 has the same natural frequency as tuning fork 1, its vibrations will continue to increase and become noticeable.

The mechanism outlined above allows for some of the energy in tuning fork 1 to be transferred to tuning fork 2. Eventually, all the energy in tuning fork 1 will dissipate, and it will stop vibrating. At this point, however, tuning fork 2 will have absorbed its maximum amount of energy from tuning fork 1 and will be vibrating at its maximum amplitude. This is the mechanism by which homeopathic remedies are believed to work.

At this point, the process will change direction, and the energy from the now vibrating tuning fork 2 will start to transfer back to tuning fork 1. This is another interesting phenomenon of resonance that has further implications.

Electromagnetic Waves

The preceding discussion has only looked at resonance in mechanical waves, specifically acoustic waves. The principles of resonance also apply to electromagnetic waves, allowing them to transfer energy in the same fashion. However, there are several differences between acoustic waves and electromagnetic waves, and some unique characteristics of electromagnetic waves make them more effective at transferring energy and information.

Direction of Medium Movement

The first difference lies in the direction of movement of the medium of the waves. Acoustic waves are longitudinal, while electromagnetic waves are trans-

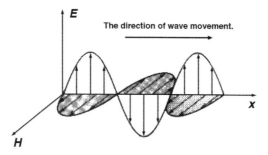

Figure 10.4. An electromagnetic wave: the E axis shows the electrical field while the H axis shows the magnetic field.

verse. In longitudinal waves, particles vibrate in a direction parallel to the direction that the wave is traveling. The medium can be considered to be alternating backward and forward. Conversely, the medium of a transverse wave moves in a direction perpendicular to the direction the wave is traveling. For a wave traveling forward, the medium can be seen as alternating left and right or up and down. Therefore, transverse waves possess a polarization plane, which is perpendicular to the direction the wave is traveling, while longitudinal waves do not. The polarization plane is the direction of oscillation. For example, if you imagine a rope being moved up and down to produce a wave, the polarization plane would be vertical. For electromagnetic waves, the situation is even more complicated. An electromagnetic wave is composed of two transverse waves, an electric wave and a magnetic wave, that are perpendicular to one another (fig. 10.4).

Visible light, an electromagnetic wave, is transverse in nature and has a polarization plane. Experts in optics have harnessed these characteristics for a variety of practical applications. One example is an instrument that uses light to measure sugar concentration in water. Sugar molecules rotate the plane of polarization, so the greater the concentration of sugar, the more the plane is rotated.

Figure 10.4 shows an electromagnetic wave's two transverse waves: the electrical wave on the E axis and the magnetic wave on the H axis. They are traveling together, and their planes of polarization are perpendicular to one another. This means that electromagnetic waves are capable of carrying more information, thereby conveying more subtlety. If we employ the analogy of music, if it were possible to create music with electromagnetic waves, the sound created would be much richer.

Speed of Wave Propagation

Electromagnetic waves are also considerably faster than acoustic waves. Acoustic waves travel at a speed of 650 to 1,000 feet per second in air and about 5,000 feet per second in metal, while electromagnetic waves travel at a speed of 1 billion feet per second. It is possible to appreciate the scale of the difference by considering the following example. If you sat in the middle seat of the front row of a theater for a broadcast concert, you might think you are the first person to hear the music. However, someone sitting next to their radio at home will hear it before you. The microphone is much closer to the singer than you are, and the whole process of transforming the acoustic waves of the music into electromagnetic waves, broadcasting them via radio to the listener's home, and the radio transforming the electromagnetic waves back into acoustic waves will take less time than it takes the acoustic waves to travel the distance to your seat.

Vacuum as a Medium

The final and perhaps greatest difference between electromagnetic waves and acoustic waves is that they use completely different media. The medium for the acoustic wave is air, which is composed of molecules. An acoustic wave moves through the combined oscillations of a multitude of air molecules. Visualizing such a mechanism is relatively straightforward if we imagine molecules as small solid balls. Acoustic wave couldn't exist without air molecules or some other material medium. In other words, acoustic waves cannot exist in a vacuum. In contrast, electromagnetic waves not only exist in vacuums, they travel around the world and across the vastness of the universe in vacuums. In other words, the vacuum is the medium for the electromagnetic wave. It is what enables them to transfer energy and information.

It is conceptually challenging, even for physicists, to accept the counterintuitive notion that the vacuum, which means nothingness, could play the role of medium in propagating waves. It is why scientists, near the end of the nineteenth century, started to search for the medium that was nothingness. They called the assumed medium "ether" and tried to study its structure and its coefficient of elasticity, and to measure its speed of movement. They failed to measure the speed of the ether and found that the speed of light, the electromagnetic wave, is somehow always the same in a vacuum, completely independent of the speed of the assumed ether. The strong characteristic of the ether, namely the constant speed of electromagnetic waves, is the fundamental axiom of Einstein's theory of relativity. In fact, the theory of relativity was only a by-product of research on the medium of electromagnetic waves. The theory of relativity is more than a hundred years old, and the original goal, to find the medium, is still not achieved—physicists have forgotten that the theory of relativity stemmed from this research, and the original question remains open.

Physicists are aware that the so-called ether is, perhaps, only vacuum, and so vacuum is the medium of electromagnetic waves. The vacuum is also the essence of the universe. Electromagnetic waves are only ripples in the vacuum, and particles are only wave packets, not the solid balls we once imagined. In other words, everything in the universe is a vacuum fluctuation, nothing more. The latest understanding of physics of the vacuum corresponds to the fundamental belief of Buddhism that "nothing is everything and everything is nothing." When high-energy physicist Fritjof Capra pointed out that the development of modern physics parallels Eastern mysticism and the basic principle of Buddhism, many conventional physicists found it highly provocative.

The widely accepted scientific theory of the Big Bang also describes the universe as starting from an explosion that was a vacuum fluctuation. In the philosophy of ancient India, the world began with a sound. In light of our understanding of physics, this sound can be interpreted as an electromagnetic vibration rather than an acoustic wave; this Indian philosophy could be regarded as an ancient version of the Big Bang theory.

In the Bible's Book of Genesis, we find that the creation of light, namely electromagnetic waves or vacuum fluctuation, happened long before the creation of the sun, stars, earth, and moon. It also means that the creation of electromagnetic waves was much earlier than solid molecules. In other words, the process of the Big Bang is a new and detailed version of the first part of Genesis.

However, I digress. The beginning of the world is not the subject of this book. What can be established is that for all the depth and complexity that acoustic music possesses, electromagnetic waves are capable of weaving a far more complex and nuanced harmony. Our bodies and other living organisms already exhibit electromagnetic music that is much richer and more charming than what is heard in concert halls. If we designate the music of concert halls as "physical" or "material" music, we might call the electromagnetic music in living systems "ethereal music" or "spirit music," since the medium of this music is the ether, the vacuum, or nothingness. Unlike acoustic music, which is limited to the atmosphere of our planet, electromagnetic music exists not only in living creatures but throughout the universe. This correlates with the idea that "heaven and human beings are united," fundamental to Classical Chinese Medicine and other holistic health systems.

Transferring Information by Means of Waves and Resonance

Nowadays, the term *information* is fashionable. Many even refer to the current era as the "information age." Information technology has been the leading industry in the last twenty years and will continue to be important in the foreseeable future.

Given this focus on information, it is quite ironic that no one can definitively state what information is. The founder of information theory, American mathematician Claude E. Shannon (1916–2001), defined information as the "degree of surprise."[2] Cofounder Norbert Wiener (1894–1964), also a mathematician, stated, "Information is neither material nor energy. Information is information."[3] These two definitions, unfortunately, do not provide any clear explanation of the essence of information, and the situation has not been clarified in recent times. In fact, information is even more intangible than energy or the vacuum.

Practically, it is possible to avoid contemplating the essential nature of information and instead think of information as messages, or signals, that can be written with a pen on paper. That said, the paper and the words are only the carriers, or the medium, of information, not the information itself. Accordingly, many other media, including a variety of different waves, can perform the role of information carrier.

In radio broadcasting, the fundamental frequency of the radio wave is referred to as the carrier wave. This is subsequently modulated by a signal that transforms it into a modulated carrier wave that is broadcast into the environment, as shown at the top of figure 10.5. Radio receivers then pick up this modulated wave and remove the carrier wave, leaving only the original signal. This procedure, shown at the bottom of figure 10.5, is called filtering.

Vocal communication employs essentially the same technique. The acoustic waves generated by the throat serve as the carrier waves, which are then modulated by the tongue, teeth, and lips to create the modulated waves we call language, which transfers information to others. The listener performs the procedure of demodulation—they receive the signals with their ears and filter out the carrier wave with the brain to understand the information in the language.

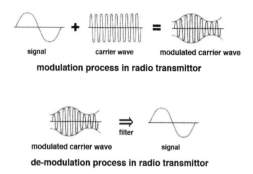

Figure 10.5. Modulation and demodulation in radio technology.

Modulation is not the only method to send information with waves. For example, there is no procedure of modulation in Morse code or in contemporary digital communication. In these cases, models based on two alternative values—"yes" or "no," dot or dash, 0 or 1—are sufficient for effective communication. The role of this form of communication in living systems, within the body or between bodies, can also be considered.

Selecting Information by Means of Resonance

In postal communication, the sender denotes the corresponding receiver using a written address on the envelope. Similarly, chemical communication in living systems employs specific chemical configurations on membrane of macromolecules, such as antibodies and hormones, to encode the address. In fixed-line phone communication, the sender and receiver are connected through wires. Nerve fibers perform the same role in the body.

Wireless communication relies on a feature of resonance to enable the sender to find the appropriate receiver. The sender and its receiver make use of the similarity or identity of their natural frequencies to ensure that the receiver is attuned to the sender's frequency. In light of this, it is instructive to reexamine the fundamental principle of homeopathic remedies: similarity, which corresponds to the resonance effect in waves.

From the discussion of resonance at the beginning of the chapter it is apparent that, in the case of acoustics, in order for tuning fork 1 to be able to transfer energy to tuning fork 2 (fig. 10.3), they must possess exactly the same fundamental frequency. This is usually achieved by making the two forks structurally

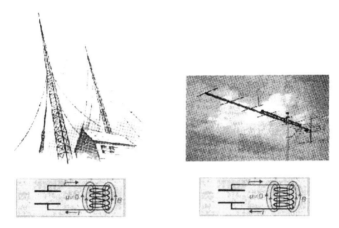

Figure 10.6. Oscillating circuits (bottom) of a sender (left) and receiver (right).

identical. The same principle applies in electromagnetic wave communication, although the sender and the receiver can have completely different appearances. The upper part of figure 10.6 shows the considerable size difference between the antennae of a sender and a receiver. Nevertheless, their internal oscillating circuits, shown at the bottom of figure 10.6, which govern their electromagnetic natural frequency, are identical.

In terms of physics, a mechanical oscillator like a tuning fork and an electronic oscillating circuit work according to similar principles. Both require the sender and receiver to have identical natural frequencies, a precondition for resonance and the key to wireless communication.

Multi-Resonance Communication

Wireless communication is one of the wonders of the modern era. Mobile phones allow us to converse with our families and friends from virtually anywhere in the world. What we currently take for granted is a privilege, a technological wonder that even emperors and kings of earlier eras would be unable even to imagine.

While the contribution made by modern electronics to our lives is already profound, the marvels produced have only harnessed a small part of the possibilities offered by resonance and wireless communication. For the most part, current mobile communication technologies make use of a single frequency, which represents only the simplest situation: single-resonance communication.

Multi-resonance communication means that both sender and receiver simultaneously employ multiple frequencies for wireless communication. This enables them to transfer much more energy and information than single-resonance communication. The recent development of mobile broadband is an example of multi-resonance put into effect. Multi-resonance communication can transfer significantly more complex information with much weaker signals than single-resonance. Also, as discussed below, it offers a possible mechanism for remote communication between minds.

In terms of multi-resonance communication, our bodies and those of many living creatures are more advanced than modern technology. Some of the phenomena in alternative forms of medicine, which exist beyond the framework of molecular biology, might be attributable to multi-resonance communication.

A Possible Mechanism of Homeopathy

According to British physicist Cyril W. Smith, there are many tiny "coherence regions" in water that allow information from the homeopathic remedy to be stored in the form of tiny standing waves.[4] The mechanism of homeopathy can be likened to the resonance between tuning forks described in figure 10.3, with the coherence region corresponding to tuning fork 2, which is able to absorb energy from tuning fork 1. Tuning fork 1 can be regarded as the pathogenic information infecting the patient. In a similar way, the homeopathy remedy can remove the harmful information from the patient. While the interaction between the tuning forks is only single-resonance and very simple compared to the complex multi-resonance interaction between a homeopathic remedy and a patient, the underlying principles are the same.

The coherence regions that Smith proposes are small sections of water. They contain an immense number of infinitesimally small oscillators, operating like tiny tuning forks, which serve to store vibrating information. This internal complexity gives it the ability to store incredibly complicated information. The tiny vibration amplitudes of these small oscillators result in very slow decay speeds for the oscillations. This allows homeopathy remedies to be stored at room temperature for years. However, for reasons that are not completely understood, temperatures above 158 degrees Fahrenheit damage the coherence regions.

The Possibility of Remote Communication through the Mind

Before undertaking any discussion about the possibility of remote mind communication, I have to confess my reluctance to voice my opinion publicly on this matter for more than fifteen years before the publication of this book. In other words, this represents the first time I have gone on record with my speculation on the matter.

From the viewpoint of theoretical physics, an interaction is occurring between minds because of the huge number of oscillators sending and receiving electromagnetic waves in each person's mind. However, as technology has not been able to create a sender or receiver with anywhere near as many oscillators, mainstream science does not believe that our minds are capable of organizing all these interactions into meaningful communication.

China has been experiencing a large-scale movement called Qigong. The name Qigong is a broad term describing an array of practices involving meditation and meditative movement that often involve some religious aspects, and some practitioners claim abilities in foretelling the future and in remote mind communication. In my opinion, any movement of such size will involve a number of charlatans, and the opinion of a university physics professor in support of the concept of remote mind communication could be employed to provide the appearance of scientific support to their propaganda. I have been deeply concerned about my words being used to this end.

It must be emphasized that the possibility of the existence of remote communication through the mind is still in the realm of theoretical reasoning, not at the stage of experimental application. The requisite precondition—that all natural frequencies in both the sender and receiver be identical—is tremendously difficult to achieve.

Near the end of World War II, vast numbers of German soldiers died in Russia. Many of their mothers knew the exact date when their sons died, and subsequent death notices confirmed the information they had received by "intuition" or the so-called "sixth sense." From the viewpoint of multi-resonance, it is understandable that there must be many identical oscillating circuits between a mother and her son—circuits that give them some special ability to communicate with very weak signals.

Experiments with ultra-weak luminescence show that during the death of a living system, the strength of bioluminescence increases several thousand times. The strength of the electromagnetic waves emitted by the dying soldiers could conceivably have increased several thousand times. This would have made it easier for their mothers to receive the signals.

In a simplified scenario, scientists at Stanford University and Princeton University have performed experiments over several decades that demonstrate that the human mind has the ability to influence random number generators in computers.[5]

What Is Belief?

The differences between Eastern and Western thought are now quite well known. Part of this difference is evident in how each system poses questions about the same thing. The typical Western question is, "What is reality?" This is a great question, and the spirit behind it led to modern science in the West. I was comprehensively educated with this spirit and its accompanying questions since childhood, even though I am Eastern.

Given this scientific background, thirty years ago, when people asked me, "Do you believe in Qigong?" I was irritated, even angry—I considered it an irrational question. At the time I thought the appropriate question was, "Do you think Qigong is real or not?"

However, I later discovered that the question "What do you believe?" is much more fundamental than "What is reality?" It is impossible for human beings to get the real picture of reality. What we usually refer to as "reality" is only a construct formed by our mind, rather than by reality itself. As Austrian physicist Ernst Mach (1838–1916) pointed out, our image of reality is only an analysis and synthesis of perceptions rather than actual reality.[6] Even so, the question about belief is not useful for the development of science, at least not in its early stages.

Approximately one hundred years ago a group of physicists, including Mach, Heisenberg, Bohr, and Schrödinger, found that it is impossible for human beings to observe the true nature of reality. What we can achieve is only analysis and organization of our feelings, and consistent interpretation of experimental results and our feelings. This phenomenon is particularly evident in the wave-particle duality of light. Depending on the design of the experiment, light will behave as either waves or particles. Belief plays a key role in the process of analyzing and organizing our experiences and feelings. In the words of Heisenberg, "What we observe is not nature itself, but nature exposed to our mode of questioning."[7] Even in the time of the New Testament, Paul said, "Faith is the substance of things hoped for, the evidence of things not seen."⁸ Two thousand years later, arguably the same understanding was achieved again and stated more clearly by Heisenberg.

Perhaps discussing such a fundamental problem in physics lies beyond the scope of this book. However, every psychologist and meditation therapist knows that belief, or faith, plays a key role in therapy. The importance of belief in medical practice necessitates discussion about its nature. Growing numbers of medical doctors have lobbied for increased research into the placebo effect, because as many as 25 percent of patients taking placebos have positive health responses.

It is possible to consider belief as a kind of resonance, or in terms of electronics, a kind of tuning procedure. Take the traditional television set as an example: if you set your television to channel 1, your set firmly believes only channel 1. It receives all the information from channel 1 while simultaneously rejecting any information from the other channels. In other words, channel 1 is the faith of your television set. If you then switch to channel 2, your television becomes a convert to the channel 2 faith. At this point, all information being beamed on channel 2 is received, while anything broadcast on any other channel, including information from its original beliefs, channel 1, is rejected.

The power of belief in is evident throughout the history of humanity. It has had the power to unite groups as well as to lead them to war. Similarly, in medicine and health care, varying beliefs and faiths can lead our minds and bodies into turmoil or into a state of coherence and harmony. Two thousand years ago, the classic Chinese medical text, the *Yellow Emperor's Canon of Medicine*, pointed to defective mental states as the source of all diseases. In modern times, however, this lesson eludes us.

CHAPTER 11

THE MYSTERIOUS AURA: FROM RELIGIOUS TO PRACTICAL

I saw another mighty angel coming down out of heaven. He was wrapped in a cloud and had rainbow round his head; his face was like the sun, and his legs were like pillars of fire. —JOHN, REVELATION 10:1

Various religions share common references to inexplicable auras around holy people (fig. 11.1). Many modern people, including me, having been educated in modern science and rationalism, find the concept of auras difficult to accept. My original opinion concerning the aura was that it was a beautiful fabricated addition to the depiction of those believed to be sacred and had nothing to do with reality. That was until I happened upon some experimental evidence that I could not disregard.

As mentioned earlier, in the 1980s, China saw the development of a considerably large Qigong movement. Like many, I was a little shocked at the scale of

Figure 11.1. The holy aura around Jesus (left) and the Buddha (right).

the movement, but I did not take it seriously. I viewed it as a religious movement brought about by large numbers of Chinese people suffering from disillusionment or a belief crisis following the death of Mao Ze-dong. People had worshipped the "great helmsman" Mao, the modern version of an emperor, almost as a living god. In my opinion their bereavement led them to turn to the many Qigong masters who could fill the role of lesser leaders and minor living gods. They served as a focal point for the worship of religiously hungry Chinese people. In honesty, after observing some of their work, I had a negative view of these Qigong masters and felt that they were liars and swindlers.

By chance, a biologist named Lu worked at an institute of Traditional Chinese Medicine adjacent to my university. In his illness-plagued youth, he had learned Qigong in order to regain his health, and at the time had initiated a very popular Qigong course at his institute. He would visit my lab from time to time and was amiable, friendly, and polite. Out of curiosity, I asked about Qigong and asked some technical questions. Later, he asked whether I could perform some scientific research into Qigong. Given my overall feeling about Qigong at the time, I immediately refused, but his amiability and sincerity led me to decline his request in a polite and subtle way. My response was something like, "Qigong might be a subject for science to consider in the future, but not at present. It is difficult to understand in terms of modern science." Of course, my real meaning was much less equivocal.

The subtlety of my rejection led Lu to misinterpret my intention. He returned the following day with a group of his friends and said, "You are a scientist and a university professor, and are open to the unknown phenomena of Qigong—this is great. If you think Qigong is a subject for future scientific consideration, why not come today and observe our course?" The situation left me unable to reject their invitation, and I accompanied him to the garden of his institute.

I have to admit that I found the experience of watching these people practice Qigong unnerving. The class began with silent meditation, then gradually proceeded into deeper and deeper states of meditation. At this point, I observed some of the participants begin to move spontaneously. Their movements became progressively more pronounced, and some of the students started to dance, sing, shout, and even roll around on the dirt in the garden. I have to concede that I found the scene repulsive.

After the course, the participants passionately discussed the experience. They felt happy and relaxed, and I began to reconsider my feeling about the class—

maybe it wasn't so bad. However, they began to talk about seeing an aura around a participant's head and about a fire among them that they had all seen. Their discussion led me to believe that they were attempting to work together to deceive me, which left me unhappy and even angry. I had been standing in the same garden and had not seen anything remotely resembling the supernatural phenomena they were describing. I did my best to suppress my anger and politely said good-bye. My personal visit so encouraged Lu that he returned to my lab and again tried to persuade me to devise a scientific experiment that would provide evidence for the existence of external Qi—life force or life energy—that they believed they were able to emit from the palms of their hands to help their patients.

My lab had just received a new double-beam spectrophotometer, which was able to simultaneously scan two test tubes to get independent absorption spectra from two specimens to discern tiny differences between them.[1] It was advanced for that time and even boasted computer processing. After some consideration, I designed an experiment to discern whether or not the Qigong practitioners were cheating. I deposited some solution of RNA—large biological molecules that are essential for all known forms of life—in two culture dishes. One acted as the control, which I placed far from us, and I put the other on my experiment table. I asked Lu to hold his right hand four inches above the dish for five minutes and to try to emit some invisible Qi into it. I then placed the RNA solution that had supposedly been treated by his external Qi into one comparison tube of the double-beam spectrophotometer, and the control solution into the instrument's other tube. I then used the spectrophotometer to scan the tubes over the whole range of visible and ultraviolet light.

I believed nothing was actually happening when they were emitting invisible Qi into the solution, and I was certain that the spectra of the two dishes would be exactly the same. This would be indicated by a horizontal straight line on the spectrophotometer's screen indicating no differences between the spectra. If, contrary to my expectations, the invisible Qi did have some effect on the structure of RNA molecules, a shift from horizontal would be easily observable in the line.

To my surprise, a sharp peak occurred in the spectrophotometer's line at a wavelength of 219 nanometers. As the design of the experiment was already quite strict, I asked Lu to invite five of his Qigong friends to repeat the experiment in order to satisfy the requirement of reproducibility. These six people would act as the treatment group. I also invited five of my colleagues at the university who had no knowledge of Qi, along with me, to form a six-person control group. The control group was asked to hold their right hands four inches above the dishes for five minutes and do nothing else.

Everyone in the treatment group managed to produce the peak at the same 219-nanometer wavelength with excellent reproducibility. Interestingly, those in the control group also produced peaks at the same wavelength. However, the average height of the treatment group's peaks was three times that of the control group. The Qigong practitioners were very happy with the results. According to their explanation, people without Qigong training, like those in our control group, also have external Qi, but it is not as strong as among those with Qigong training.

Once the results had been analyzed, the difference between the two groups was found to be statistically significant. In other words, we had performed a very elegant and reliable experiment. Unfortunately, I have to admit that I did not dare to publish the results. I was concerned that the nature of this experiment would greatly offend the old professor who was my boss at that time. Nevertheless, Lu and his Qigong colleagues were so pleased that, in addition to showing the results to their friends at every opportunity, they offered me a complimentary advanced Qigong course. I accepted their offer, partially because it was being held during summer vacation in a Buddhist temple surrounded by beautiful tranquil mountains.

While I found the course to be quite odd, I did my best to be, or at least to appear to be, a good traditional Confucian Chinese student by obeying any instruction the teacher gave without asking why. On the first day, my teacher asked me to sit and maintain a posture like holding a ball with two hands, one above another, palm to palm, for two hours. The ball was invisible and untouchable, at least to me. The teacher proudly said it was a Qi ball. I harbored the uncomfortable feeling that the situation was akin to Hans Christian Andersen's "The Emperor's New Clothes." This feeling arose repeatedly during the Qigong course, but I was not brave enough to voice it.

The Qigong course was an advanced one; except for me, all of the participants were Qigong masters. They listened to the lecture with great interest and diligently took notes. I was most unimpressed because the theory being espoused was completely at odds with the scientific theories I had been educated in since childhood. I could not imagine what kind of notes one might take from such crazy nonsensical discourse, and was curious to find out.

The Mysterious Aura: From Religious to Practical

After class one day, when the other students had left, I noticed that someone had forgotten their notebook. I furtively took the opportunity to see what was inside. In addition to some notes on one of the crazy theories, the student had also written some of his personal observations. He had seen a bright golden aura around the teacher giving lectures. I had been in the same classroom, watching the same teacher, and had seen nothing like this, but I was mindful that this was recorded in a personal notebook, not for propaganda or deception. From that point, the question of auras persisted in my mind for ten years, until I had traveled to Europe, learned some related material, and observed other evidence.

The first place I encountered experimental evidence concerning auras was in Kaiserslautern, Germany, in 1991. The city was home to an international institute of biophysics, where people measured the extremely weak luminescence emitted by living systems, such as insects and small plants, in completely dark chambers (fig. 11.2). Their instruments detected emissions in the range of visible light and ultraviolet radiation. The spontaneous luminescence of living systems is very weak, in the range of 320 to 650 photons per second per square inch, a little above the quantum background noise, which is around 130 photons per second per square inch. When electromagnetic waves are measured as particles, they are called photons. The variation in photons produced by quantum effects is always present. The rate of 320 to 650 photons represents very weak light, comparable to that received from a candle ten miles away. It lies beneath the visible threshold of even our dark-adjusted eyes.

The institute's studies found that at the moment of death, spontaneous luminescence would increase several thousand times. Consequently, the spontaneous luminescence would be as strong as the light of several thousand candles ten miles away. While this is still weak, it would be visible with the naked eye in a dark place. Given the feasibility of viewing this luminescence, it is not too difficult to envision how people in ancient times perceived the aura with their eyes. For example, someone keeping vigil at the deathbed of a person in a dark cave would be able to discern an aura at the moment of death.

This phenomenon might not be limited to dying people. For individuals in relatively unique physiological states, such as a very high fever, or unique mental states, the spontaneous luminescence could also greatly increase. These individuals might also have heightened visual sensitivity.

While spontaneous luminescence from living systems is extremely faint, the photographic evidence from the experiment, shown in plate 14 in the color plate section, among many others, provides serious, scientific, and powerful evidence of the objective existence of a kind of aura.

The second place I witnessed scientific evidence of auras was at the Russian Academy of Sciences in Moscow in 1993, two years after the Soviet Union broke apart. I visited the lab of Eduard Godik, who was an authority on remote monitoring by spy satellites, an important military technology that had been awash in government research grants. His lab was so lavishly equipped—sensitive photographic equipment, the best infrared camera, and the best microwave detection system—that I must admit I was envious. The sudden collapse of the Soviet Union saw the lab completely lose its financing and led Godik to seek support from other sources. Samsung, the giant Korean electronics company, suggested using the facilities for medical research, and provided some funding.

Imagine the level of detail that can be achieved when using remote-sensing equipment designed for observing people at close proximity from hundreds of miles away. Godik's group immediately saw patterns of light around human bodies (plate 4 in the color plate section) that matched the descriptions of religious figures thousands of years ago.

Biologists have known for years that snakes can discern a mouse in a completely dark room by means of an infrared sense organ near the nose. Biologists also know that bees are able to sense ultraviolet light from the sun, allowing them to see the sun even when thick clouds obscure it. In this way, snakes and bees are much more powerful than humans. In the terminology of physics, they can perceive a much wider range of electromagnetic frequencies than we can.

Conversely, some people exhibit color blindness. This can range from those who are unable to distinguish between red and green, called anomalopia, to those who see no color at all, called achromatopsia. People with achromatopsia perceive the world much like we see a black-and-white movie. Patients suffering from some forms of color blindness do not see some hues at all and perceive them as black. These patients have reduced visual ability in a range of frequencies. In other words, they perceive a smaller range of frequencies than most people.

Conditions where people have reduced visual ability compared to us are easily understood, but cases where people possess stronger visual ability, who can perceive a greater range of frequencies than normal, would be misunderstood. Someone claiming to see things that we do not would be assessed as suffering from visual hallucinations. Even worse, as with my experience in the first Qigong course, we might conclude that they are lying, trying to scam us, or even suffering from a serious mental disorder.

Fortunately, people today have more tolerance. Some psychiatrists, including Andrew Powell, chair of the Spirituality and Psychiatry Special Interest Group of the Royal College of Psychiatrists in the United Kingdom, have already found that some diagnosed psychotics are not really suffering from mental disorders but are instead in a unique physiological state. Powell even suggests that such a state might be caused by receiving information from "another world."[2]

From the viewpoint of physics, it might be unnecessary to introduce the notion of another world, as it can be explained by information being transmitted in the form of invisible electromagnetic waves. Extraordinary physiological states could potentially heighten an individual's sensitivity, enabling them to see beyond the normal visible-light range of electromagnetic waves. In this scenario, the argument can be made that while they are suffering from a physical disorder induced by some invisible physical factor, they have no psychological ailment. Of course, this is only a hypothesis that needs more experimental proof.

The phenomenon of water dowsing might serve as an example to illustrate more clearly the situation of heightened sensitivity in humans. Dowsing is an ancient technique employed in Europe to search for underground water or metals by using a Y-shaped stick or rod. Some modern scientists postulate that the working principle of dowsing is to detect a sharp change in the earth's magnetic field where there is underground water or metal. However, ordinary people do not have the ability to detect such weak variations in magnetic fields.

Cyril W. Smith, a physicist and electronic engineer, has spent much of his life studying the electromagnetic aspect of human bodies. In addition to his serious scientific research on frequency reactions in the body, he also possesses the special ability of dowsing. During a private conversation, he told me that this ability actually represents an infirmity; his body is unable to compensate for sharp changes in electromagnetic fields. He also said that dowsers are more prone to heart attacks than regular people. He never refers to himself as having special abilities, instead labeling himself a "patient of oversensitivity." Another example of oversensitivity to variations in electromagnetic fields are people who suffer from periodic headaches in accordance with the phases of the moon.

It is likely that people in deep meditative states have increased sensitivity to acoustic and electromagnetic waves. This is a possible explanation as to why people in ancient times were aware of the existence of the invisible aura. There was much less background noise and far fewer distractions than in the modern world. Traditional culture allowed for a more relaxed, less stressful, less fast-paced life, which would facilitate deeper meditation.

The aura referred to in ancient times might also include the chemical aura. We expel large numbers of molecules from our bodies. With their highly developed sense of smell, dogs can not only distinguish people from each other but even gain insight into a person's health. Some dogs are trained to detect when their diabetic owner's blood sugar is outside an acceptable range, and research is being done into whether dogs can be trained to detect cancer in their owners.[3] This book will focus on the electromagnetic aura and its potential applications.

The Practical Application of Auras in Medicine

It can be argued that the first practical application of auras involves the technique of high-frequency, high-voltage photography. In figure 11.4, an aura that appears to frame the hand is clearly evident. If we focus in on two of the fingers, at right, we can see that the aura resembles the corona—the plasma that surrounds the visible surface of the sun.

High-frequency photography was invented in 1926 in the Soviet Union by Semyon D. and Valentina K. Kirlian, and is widely referred to as Kirlian photography. When people are charged by very high-voltage current, in excess of 3,000 volts, for a very short time, less than 10 milliseconds, they will not be hurt but instead become highly charged with electricity. When the charging current ends, a discharge process occurs, and a discharge pattern can easily be recorded by photographic plates in a dark room.

The invention of Kirlian photography inspired doctors and scientists worldwide, including Semyon Kirlian, to attempt to apply this technique to medical practice. He worked in a botanical institute and found that the shape and color of an aura is closely related to the physiological state of a plant. At the time, scientists and medical doctors expected Kirlian photography to be potentially useful in distinguishing various physiological, pathological, and even psychological states. Many others proposed that Kirlian photography might bear some relationship with the acupuncture system.

Attempts at practical medical application of Kirlian photography were not successful until the 1970s, when German medical practitioner Peter Mandel developed the method of Kirlian photography exhibited in figure 11.5. The image

The Mysterious Aura: From Religious to Practical

Figure 11.4. Aura around a hand and two fingers, seen through Kirlian photography.

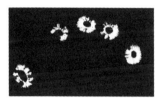

Figure 11.5. Kirlian photograph for practical diagnosis.

shows ring-like auras surrounding the tips of the fingers of the right hand. It is evident that there is a breach on the left side of the index finger. Interestingly, whenever a patient has loose bowels, there is always a gap present at this location in a Kirlian photograph. This observation is consistent with acupuncture theory, as the Large Intestine Meridian begins at this location.

While it is arguable that level of precision achievable through Kirlian photography–based diagnosis does not match that of Western medicine, it does show the auras carry some information about the body. It also supports the argument that the aura is related to the acupuncture system. Further evidence was discussed in chapter 5, where a continuous high-frequency, high-voltage photograph taken in a dark room clearly shows the track of an acu-meridian (fig. 5.9).

It must be pointed out that unlike the spontaneous light emission evident in figure 11.2, or the infrared thermograph in figure 11.3, the aura in the Kirlian photography is not a real aura, but an artificial one. In terms of physics, Kirlian photography produces only a discharge pattern, but it is very closely related to the distribution of the electromagnetic fields inside the body. Because the acupuncture system is a simplified representation of the distribution of the internal dissipative structure of electromagnetic fields in the body, the two have a close relationship.

The Aura inside Our Bodies

The preceding discussion illustrates that the invisible dissipative structure of electromagnetic fields inside the body can also be regarded as an aura inside the body. This internal aura is always intermingled with the physical chemical body and would be easily damaged by the intrusion of a scalpel. As such, it is challenging to observe and has consequently been ignored by scientists for hundreds of years, and almost completely forgotten.

In terms of understanding this internal aura, ancient cultures were much more knowledgeable than modern people, even scientists. How people in ancient times were able to find something as elusive as the acupuncture system in Traditional Chinese Medicine, or the chakras in traditional Indian medicine, is lost. Both of these systems refer to auras inside our bodies; the acupuncture system is beneath the skin, while the chakras are along the central vertical axis of the body.

According to the theory of traditional Indian medicine, each of the seven major chakras (plate 15 in color plate section) has its own name, function, and color. It is worth noting that the word *chakra* means "light ring" in Sanskrit. Variations in the color, shape, and direction of spin of these light rings indicate psychological, physiological, and even pathological changes in the person. In other words, the chakra system is closely related to the mind-body system.

From the viewpoint of Western biologists and medical doctors, these seven rings have never been discovered in anatomical research, and so chakras are viewed as nothing more than appealing folklore. From the viewpoint of physics, however, there must be focus centers of waves, both acoustic and electromagnetic, along the central line of the body. This is a simple outcome of the property of reflection inherent in all waves (fig. 11.7).

In Classical Chinese Medicine there exists the puzzling Triple Burner Meridian. Other meridians, such as the Lung Meridian, Stomach Meridian, and Kidney Meridian, correspond to unambiguous solid organs in anatomy. But for a long time, there was no concept of what organs the Triple Burner Meridian corresponded to. In 2012, Yetao Gao, a medical professor in Nanjing, pointed out that the triple burner corresponds to the three hollow organs, also referred to as the three cavity organs, namely, the cranial cavity, the chest cavity, and the abdominal cavity, as shown on the right side of figure 11.7.

Interestingly, Gao made this intriguing discovery through literature research and clinical experience. He did not have a background in physics and was not

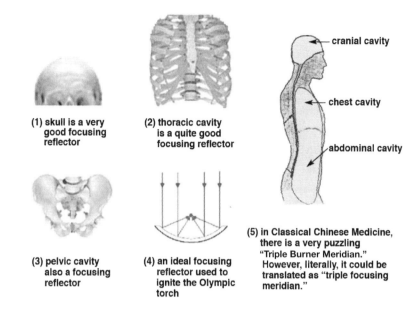

Figure 11.7. Reflectors along the central line of the body.

aware of the relationship between cavities and the focusing of waves. When I read about his work I was amazed by the wisdom of the ancients, who not only discovered the three biggest focusing reflectors in the human body, but also chose appropriate words to describe them.

The property of wave reflection also dictates that there must also be many smaller focus centers caused by other curves in anatomic structures.

Detection and Data Analysis of Auras

From the viewpoint of theoretical physics, the existence of the external and internal aura is consistent with the chakras and the acupuncture system, but the practical problems of detection and data analysis remain. Technically it is not too difficult to detect the external aura; it can be detected in living systems by means of existing instruments, including sensitive infrared cameras, microwave sensors, and photomultipliers, which are extremely sensitive light-detecting vacuum tubes capable of detecting individual photons. Detecting aspects of the internal aura, such as chakras and acupuncture meridians in living systems, is more challenging, because, as discussed earlier, the aura is intermingled with the chemical body, and significant disturbance to the aura is caused by the insertion of any detector or the surgery required for insertion.

Fortunately, we can detect the heterogeneous distribution of body conductivity on the surface of the skin. This is particularly useful, because this conductivity is proportional to the distribution of the internal electromagnetic field inside the body. In other words, the heterogeneous distribution of body conductivity reflects the energy distribution inside the body, namely the internal aura. Unfortunately, body conductivity measurement, or Kirlian photography, only enables us to detect the outer surface of the internal aura. We still do not have methods to detect the chakras system, even indirectly; their practical detection remains an open problem for technology.

The biggest challenge in studying the aura is data analysis. Unlike molecules, which can be counted individually, the aura is a composite of a practically infinite number of inseparable electromagnetic waves. Therefore, analysis of even the simplest aura measurements is already far beyond the ability of the most advanced methods of calculus and other mathematics. Brilliant mathematicians have already developed new branches of mathematics that can deal with complex systems composed of infinite elements. Equipped with these methods, it is possible to start to quantitatively study the information gathered from measuring the aura. The final part of this book will introduce some practical methods for calculating the degree of coherence or harmony in aura measurements.

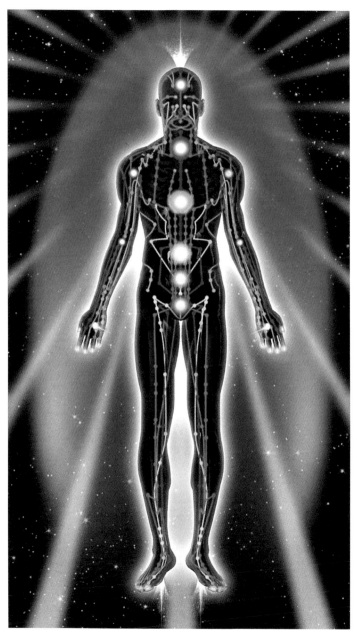

Plate 1. The Electromagnetic body. Both the meridian system in Chinese Medicine and the Chakra system in Indian Medicine can be explained in terms of the electromagnetic body.

Plate 2. Glucose molecule. A simplified ball and stick representation of a glucose molecule. Even though Physics has discovered the more complex interactions in molecular bonding most still visualize molecules in terms of this type of model. (Chapter 2, figure 2.2)

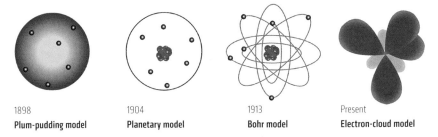

1898	1904	1913	Present
Plum-pudding model	**Planetary model**	**Bohr model**	**Electron-cloud model**

Plate 3. Different models of the atom. Advances in our understanding have forced o ur model of the atom to evolve over time. (Chapter 2)

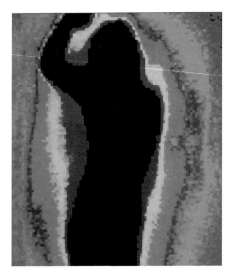

Plate 4. Infrared Aura. As photographed with highly sensitive scanning equipment. (Chapter 3, figure 3.2 and figure 11.3)

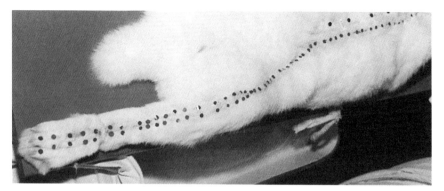

Plate 5. Meridian on a rabbit. Points of lower resistance (blue) and higher sound intensity (red) on the skin of a rabbit correlate closely with meridian locations. (Chapter 5, figure 5.11)

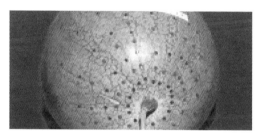

Plate 6. Meridian points on a watermelon. Points of lower resistance (blue) and higher sound intensity (red) on the surface of a watermelon. (Chapter 5, figure 5.12)

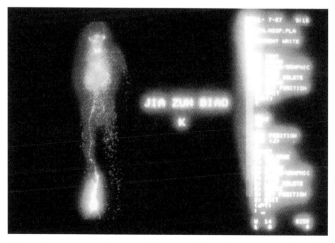

Plate 7. Isotope tracer showing kidney meridian. Some compounds will also follow the path indicated by the meridians. (Chapter 5, figure 5.13)

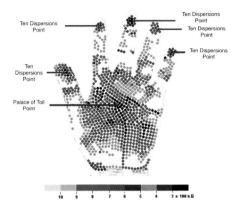

Plate 8. Typical Conductance profile of a hand. Points of consistent high conductivity correlate with the location of acupuncture points.(Chapter 6, figure 6.3)

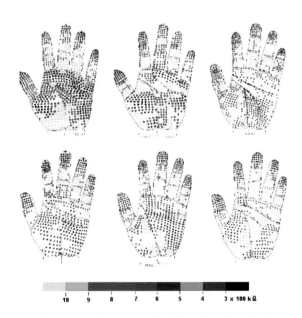

Plate 9. Conductance profile of the palms for individuals in various physiological states. Changes in a person's physiological state is reflected in different patterns of conductance. (Chapter 6, figure 6.3)

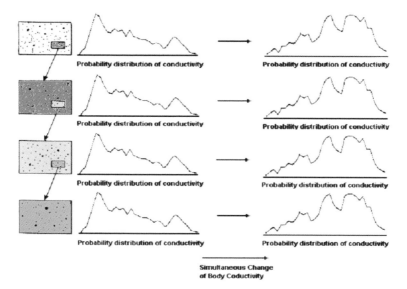

Plate 10. The holographic nature of electromagnetic field changes in the body. A change of state in one part of the body is present throughout the body. (Chapter 6, figure 6.9)

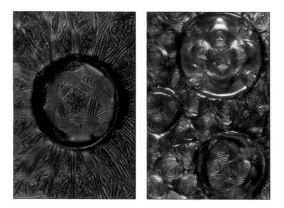

Plate 11. Boundary conditions and interference patterns of standing waves. The placement of small rings in the water greater changes the standing waves generated in the system. (Chapter 8, figure 8.8 [file name is 8.10])

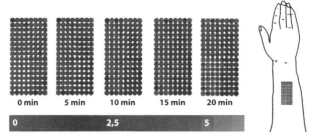

Plate 12. Currents measured at different locations on the skin show the change in interference patterns during needling. It takes 10–20 minutes for an interference pattern to reach a new stable state after a change in boundary conditions. (Chapter 8, figure 8.11 [not 8.14])

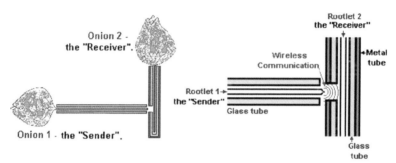

Plate 13. Wireless communication between two onions. Transmissions from the first onion influenced the rate of growth in the second onion. (Chapter 9, figure 9.4)

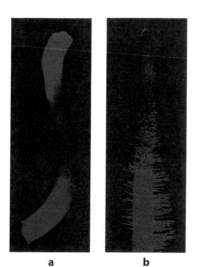

Plate 14. Spontaneous light emission in plants of: (a) an oats root; (b) a maize shoot. Electromagnetic waves in the visible range being emitted by plants. (Chapter 11, figure 11.2)

Plate 15. The seven chakras and their colors. (Chapter 11, figure 11.6)

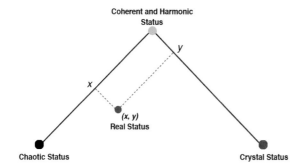

Plate 16. The coherence pyramid, a condensed infinite dimensional space. (Chapter 14, figure 14.12)

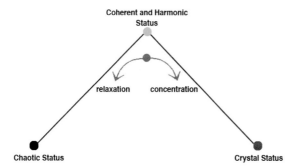

Plate 17. Coherence is present when a person constantly switches between concentration and relaxation. (Chapter 15, figure 15.1)

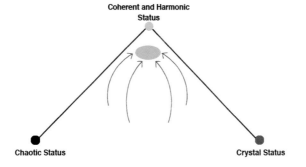

Plate 18. The green area is the normal state, and attracts the state of a person to automatically regain coherence. (Chapter 15, figure 15.2)

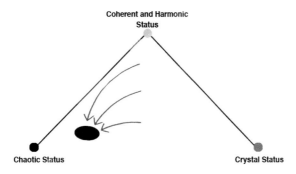

Plate 19. An incorrect attractor near the chaotic state. (Chapter 15, figure 15.3)

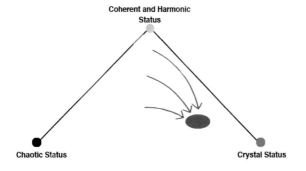

Plate 20. An incorrect attractor near the crystal state. (Chapter 15, figure 15.4)

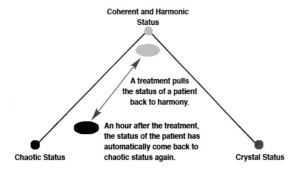

Plate 21. The tug-of-war phenomenon in treating chronic diseases: a treatment pulls the patient back to coherence (green area), but an hour later, the patient has returned to the chaotic state (black area). (Chapter 15, figure 15.5)

PART 5

MEASURING COHERENCE

CHAPTER 12

FACING COMPLEX SYSTEMS: THE END OF REDUCTIONISM

In modern physics, one has now divided the world not into different groups of objects but into different groups of connections.... What can be distinguished is the kind of connection which is primarily important in a certain phenomenon.... The world thus appears as a complicated tissue of events, in which connections of different kinds alternate or overlap or combine and thereby determine the texture of the whole. —WERNER HEISENBERG

The End of Reductionism

Biology, which forms the basis of Western medicine, has been undeniably successful and has made great contributions to the treatment of illness. At the heart of modern biology lies reductionism, a philosophical process that involves dividing the whole into progressively smaller parts in order to reduce the factors being considered. This very clever and successful approach to scientific research in physics, chemistry, biology, and medicine has been the doctrine of modern science since its inception. Its dominance has seen it established as a standardized way of thinking in scientific training, and it is the standard of the modern system of formal education.

The development of Western medicine naturally followed, and continues to follow, the path of reductionism. Thus anatomy leads to histology, histology to cytology, and cytology to molecular biology. Molecular biology, lying at the end of the reductionist pathway, is considered the highest level of biology. Many scientists, particularly biologists, believe that the solutions to all problems in medicine lie in greater knowledge of every molecule in the human body.

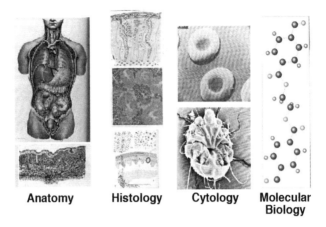

Figure 12.1. The development of biology and medicine along the path of reductionism.

The question is whether reductionism is the only approach for the future development of science and medicine, and whether reductionism is sufficient to study music, living organisms, and other complex systems. The situation of biological and medical research nowadays is somewhat like trying to study how an orchestra (fig. 12.2) would approach playing a beautiful and harmonious symphony with a reductionist approach. The first step would be to divide the orchestra into its three major constituents: the stringed instruments, the percussion instruments, and the wind instruments. This mirrors the way the human body is divided into the respiratory system, the circulatory system, the digestive system, and the nervous system. Reductionism can be continued by dividing the stringed instruments into harps, double basses, cellos, violas, and violins, similar to the digestive system being divided into the mouth, stomach, intestines, and so on. If the violins were then further divided into first violin, second violin, and third violin, the level of reduction would have reached the basic organs of the orchestra—the individual musical instruments.

In order to maintain the good health of every instrument in the orchestra, its individual components must be considered. In the case of the first violin, for example, each component, including the bow, the body, the neck, the bridge, and the strings, must be individually examined to ensure they are functioning properly.

Taking the process of reductionism one step farther, in addition to considering the components of the violin, the materials that compose them must also be

Facing Complex Systems: The End of Reductionism

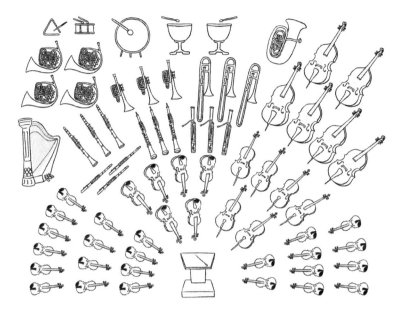

Figure 12.2. Studying a symphony using reductionist thinking.

investigated. Thus the way in which the molecules combine to constitute these materials must be assessed. The wood of the body is a decisive factor in making a good violin, so molecular research into the materials of the violin is as important to the orchestra as research into molecular biology is to medicine.

This type of detailed scientific research would inevitably be a valid approach. Research at the molecular level would definitely have the potential to improve the quality of every instrument in the orchestra. However, it is readily apparent that knowledge of all the molecules in all the instruments in an orchestra would not guarantee the quality of a symphony. The knowledge of the anatomic structure of a musical instrument and the inherent molecules is only a necessary condition to play good music. It is not a sufficient condition.

In order to perform a successful symphony, every instrument requires a skilled musician to play the instrument with the appropriate rhythm, which has nothing to do with knowledge of the molecules. All musicians in the orchestra have to exhibit a high degree of coordination to achieve the correct dynamic space structure of all the vibrations involved. This cooperation cannot be studied with reductionist thinking: holistic thinking is required to study a symphony.

Nowadays, increasing numbers of doctors are introducing some aspect of holistic forms of medicine into their practice. This implies that more medical doctors are becoming aware of the limitations of reductionist thinking and are aware that molecular biology lies at the end of the road of reductionism in medical research.

Despite the growing prominence of holistic approaches to health care, most biologists remain engaged in reductionist research because it appears that they have no other choice. Holism is a great ideal with positive connotations, but such idealistic notions appear at odds with practical scientific research. In reality, the different cultures in our world possess different ways of thinking. Within these different modes exists the means to study a complex system using practical methods and practical mathematics. To illustrate these new approaches to scientific research of complex systems, let us consider the major differences in thinking between what may be considered the Yellow River culture and the Nile River culture.

Different Structures of Thinking

Around seven thousand years ago, there were approximately twenty different major cultures. Most developed along rivers, and most disappeared. Today, only two of these ancient cultures still have a major influence on present-day styles of thinking: The culture that developed along the Yellow River in China, and the one that developed along the Nile River in Egypt.

It is interesting to compare the different ways of thinking in the two cultures. The image at right in figure 12.3 depicts the pyramids in Egypt, which are symbolic of the Nile River culture. It disappeared with the rise of Muslim empire, but its basic framework of thinking is exhibited in the stories about Moses, who was born, raised, and educated in Egypt. This culture mixed with Greek culture and became the basis of Western thinking, and thus the basis of modern science and Western medicine.

The image at left in figure 12.3 depicts the Great Wall of China, representative of the Yellow River culture. Geographic impediments like the Great Wall isolated its development from the Nile Culture until the twentieth century, and consequently it is possible to discern fundamental differences between Eastern and Western thinking.

The traditional symbols for medicine in these two cultures are indicative of the different modes of thought. The image at left in figure 12.4 depicts the

Facing Complex Systems: The End of Reductionism

Figure 12.3. Two different ways of thinking: Yellow River culture and Nile River culture.

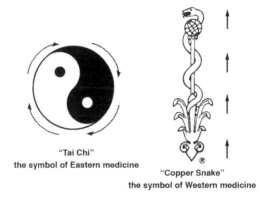

Figure 12.4. Basic structures of Eastern and Western thinking.

tai-chi, which is central to the I Ching, the "Classic of Transformations" text written four thousand years ago. The image at right in figure 12.4 is the copper snake in the 3,500-year-old story of Exodus in the Bible. The differences in the structure of medical thinking can be seen in these two images.

The central idea of the tai-chi is the dynamic balance between yin, depicted in black, and yang, in white. It represents a permanently rotating interaction of the two; in terms of modern physics, this can be regarded as a form of dynamic oscillation. The basic structure of thinking in Traditional Chinese Medicine concerns this dynamic balance. Illness in a patient stems from some sort of imbalance, and the role of medical treatment is to help the patient regain dynamic balance.

Invisible Rainbow

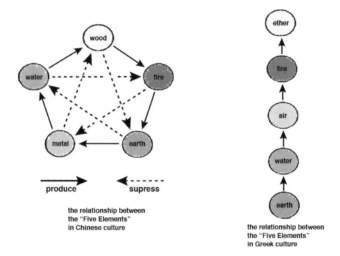

Figure 12.5. Two five-elements theories, from ancient China (left) and ancient Greece (right).

In the story of Exodus, the copper snake was used by Moses to heal patients. The symbol represents the use of powerful methods to conquer disease. This is the basic idea of Western medicine.

Further insight into the difference between Eastern and Western thinking can be discerned by comparing the five elements theory of ancient China with the five elements theory of ancient Greece (fig. 12.5). In the Greek five elements theory, the structure of the elements is linear and proceeds vertically, like the structure of the copper snake. This can be contrasted with the nonlinear structure of China's five elements theory, where two sets of relationships, "producing" and "suppressing," interact in a network that possesses neither beginning nor end.

The structure of the mathematics underlying the two systems of thought gives further detail as to their inherent differences. The left image in figure 12.6 is a representation of the I Ching, the technical development of the tai-chi, which lies at the center of the diagram.

The I Ching develops the idea of dynamic balance by describing it with digitized data, using a longer stick (———) and two shorter sticks (— —). These can be thought of as corresponding to the binary numbers 1 and 0, respectively. Therefore, the tai-chi can be further developed using these single-digit binary numbers. This extrapolates the 1 (representing yang) and the 0 (yin) in the cen-

Facing Complex Systems: The End of Reductionism

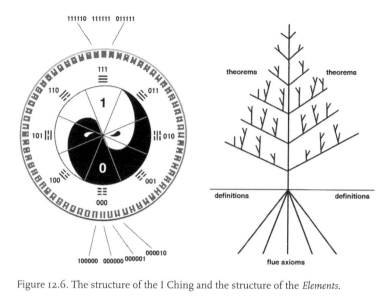

Figure 12.6. The structure of the I Ching and the structure of the *Elements*.

ter of the I Ching to the possible combinations of the three binary numbers: 000, 001, 010, 011, 111, 110, 101 and 100. In the third concentric level of the I Ching, the three-digit combinations can be further compounded into all the possible six-digit combinations: 000000, 000001, 000010, and so on, to 111111, and finally to 100000. Additional levels can be developed using the same principle, but the basic structure remains a circular network. Again, there is no beginning or end to this structure.

The diagram on the right in figure 12.6 depicts the structure of the axiomatic system, first presented by the Greek mathematician Euclid around 300 BCE in his classical text *Elements*, which sets out the rules of Euclidean geometry. The axiomatic system provides the basic standardized model of any modern scientific theory. Its structure is like a tree, starting with the roots at the bottom and ending with the smallest branches. Like the structure of the copper snake or the five elements theory of ancient Greece, this structure always develops upward.

A series of axioms form the basis and foundations of a theory. An axiom usually arises from an observation or experience that cannot be inferred from more fundamental knowledge. In other words, an axiom is a boundary between the unknown world and a collection of empirical knowledge, which can be experimentally verified but cannot be arrived at through logical inference, and therefore lies at the frontiers of our rational knowledge.

The axiomatic system is accompanied by a definition system that assigns a clear and fixed meaning to all terminology. This allows us to avoid misunderstanding in the procedure of rational inference. Together with the labor of countless mathematicians over many generations, this rigorous system has allowed many theorems to be generated. The axioms and definitions form the roots of a big tree, while the many theorems are the trunk, twigs and flourishing leaves. The elegance of the tree structure meant it became the standard model for modern scientific theory. Euclid's example of reasoning equipped future generations of scientists with the rigorous means to achieve a magnificent accumulation of knowledge.

When the treelike structure of the axiomatic system is compared to the circular structure of the I Ching, it becomes evident that the treelike model is not the only way to structure theories. However, the strong influence of its style of thinking results in almost all modern scientific theories being linear. The 1970s saw a growing realization within the scientific community of the limitations of linear thinking, which led to the study of "nonlinear problems." In terms of linearity, the structures of the Egyptian pyramids, the medical symbol of the copper snake, the Greek five elements, and Euclidean geometry are all linear structures: they possess clear beginnings and ends.

Interestingly enough, some recent considerations of the Big Bang theory move away from linearity to some degree, with the supposition that once its expansion concludes, the universe will contract with all it contains into a black hole that will birth a new Big Bang. In other words, the end of this world would be the beginning of a new one. This is a circular theory, without beginning or end, rather than a linear development with a clear beginning and end.

The study of complex systems that cannot be divided into small parts has also led many physicists to become aware of the limits of reductionism. This awareness is tied to the consideration that the whole is bigger than the sum of its parts.

Synergetics and the Study of Complex Systems

The last three decades have seen movement toward the scientific study of complex systems. Hermann Haken (born 1927) of Freiburg University in Germany can be considered one of the pioneers in studying complex systems seriously and rigorously in accordance with the standards of modern science. He proposed the idea of synergetics to explain the formation and self-organization of patterns and structures in open systems that are far from an equilibrium state. His approach divided a system into many subsystems that work together cooperatively and with coordination. Using this method, he identified the long-term and short-term parameters in a system to explain its behavior.

This area of study has continued to grow, and currently many groups in different countries are engaged in the study of complex systems. These include Hans-Jürgen Stöckmann and Bruno Eckhardt of Marburg University, who concentrate on the study of complex systems and chaotic waves, as well as at least eight special institutes located around the world. Today, more and more talented physicists are venturing into this challenging research area.

The study of economics represents another field where complex systems confront us. Economics has become increasingly important since the end of World War II, leading to the establishment of the Nobel Prize for economics in 1969. The purpose of studying economics is to discern the rules or laws of highly complicated economic systems.

In studying complex systems, the limitation and weakness of reductionism increasingly become obvious. In order to continue to advance our understanding, new mathematics and new ways of thinking have to be introduced.

The body and mind—the object of medicine, biology, and psychology—is obviously a complex system, perhaps the most complicated system facing us. The reintroduction of some holistic forms of medicine from various ancient cultures is recent, and the focus has been on developing the effectiveness of the techniques and trying to explain their function in terms of modern science. This may require scientists to explore different modes of thinking from various cultures to develop modern science to meet this challenge.

Combining Different Ways of Thinking

It must be admitted that it would have been impossible to develop, or even initiate, modern science without reductionism and linear thinking. This is perhaps why modern science developed in the West and not in the East. Science had to develop from simple systems to complex systems, from the study of linear problems to the study of nonlinear problems.

As things currently stand, the medical profession is in the position of having embraced numerous ancient Eastern therapies as well as other holistic treatments without knowledge of the underlying mechanics. This lack of understanding is not their fault; considerable effort has been made to explain these therapies within Western medicine's framework. Medical professionals have also introduced many modern techniques, such as electronic measurement, electric stimulation, soft-laser stimulation, extra-weak microwave stimulation, and colored-light stimulation to augment these ancient therapeutic methods with modern technology.

The underlying challenge is how to combine Eastern and Western thinking to extend the ability of modern science while maintaining its worthy tradition of strictness. This is the main topic of the final part of this book, which introduces the coherence pyramid, which represents this combination of Eastern and Western thinking.

Integration of Tripartite Thinking: East, South, and West

So reductionism is not the only mode for scientific development, and linear thinking is not the only approach to studying a system. In addition, shielding and isolation are even more limiting. While these methods have a long history of successful scientific pursuits, their limitations are fully exposed when confronted with scientific research into complex systems and holistic medicine.

Both Eastern and Western thinking have their advantages and disadvantages. In the present-day globalized world, a successful fusion of their respective advantages is required to find a new way to further develop modern science, and it is a very important issue for today's scientists to consider.

In fact, with the current degree of globalization, we should not only consider the different mind-sets of East and West but also look south for an additional mode of thinking that should also be considered. Differences in thinking are a very complex issue, and this book will consider it mainly from the viewpoint of medicine.

Figure 12.7 contains three images that depict completely different views of the human body, from the viewpoints of the ancient Chinese, Greeks, and Indians, respectively. These three different views of the human body evolved into three distinct medical models.

From the perspective of the ancient Greeks, the skeleton is the most important part of the body (middle image in fig. 12.7). Naturally, without the skeleton, the body would be an immobile blob of flesh, and so correct bone structure is a fundamental requirement for health. This viewpoint contributes to the popularity of orthopedic treatments in the West, which can assist with the many ailments that originate from even minor bone dislocations.

Facing Complex Systems: The End of Reductionism

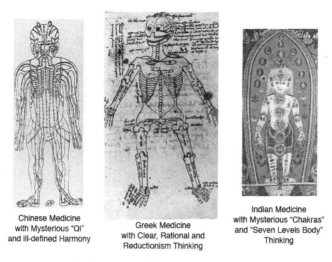

Figure 12.7. Medical mind-sets of ancient China, Greece, and India.

The skeleton of an individual is not readily apparent; we never see the skeleton of a living person. Skeletons can only be visualized as a construct in our minds. It is not too difficult to see the skeleton of a corpse, and it was from their study that medicine developed in ancient Greece. Consequently, Western doctors, deeply influenced by the culture of ancient Greece, attach great importance to anatomy and make it the basis of medical education. This is not the case in Chinese medicine.

Modern Western medicine developed rapidly in the United Kingdom, and British doctors were highly motivated to dissect corpses in order to learn more about the body. As the church expressly forbade this practice, the economic reality that scarcity increases value led to profitable clandestine dealings in black-market corpses.

While visualizing the skeleton construct that existed in the minds of the ancient Greeks posed its challenges, these paled in comparison to those facing the Chinese with their the meridian-centered mental construct. A skeleton can be seen inside a dead body, but it is impossible to examine a set of meridians. This is a major reason those disbelieving of Chinese medicine hold that meridians are a fantasy invented by the ancient Chinese.

As if these imperceptible meridians weren't challenging enough for scientific research, scientists also have to contend with the mystifying Qi, which circulates in the meridians and all around us. The challenge of studying Qi notwithstanding, perhaps the most enigmatic Chinese medical concept is "harmony," which has no tangible definition at all.

The Indian medical system makes even greater departures from the material world than the Chinese. Similar to meridians, which seem to exist only as a mental construct, traditional Indian medicine envisions chakras in the mental image of the body. It also states that people have seven layers of body, of which only the first layer is visible. While traditional Indian medicine names, describes, and analyzes the other six layers that exist beyond the chemical body, from the viewpoint of modern science, how can they be examined?

More broadly, Western medicine and modern science developed within the Greek framework in a process with no influence from Chinese or Indian thinking. In the last few decades, models of Chinese and Indian medicine have begun to pose challenges to both Western medicine and modern science. This represents a process of cultural fusion, but from a short-term perspective, the experience is a somewhat painful collision. In the long-term, this fusion will produce a new and more beautiful culture. Medicine and science, as part of this culture, will greatly benefit from this integration.

The Past, Present, and Future of the Three Cultural Circles

The past two or three thousand years have seen European culture experience a long process of cultural fusion that has been greatly beneficial. The process led European culture to occupy its currently dominant position. A brief summary of this history can help to develop a better understanding of modern science and medicine.

The relative absence of transportation technology in ancient times saw a wide variety of distinct cultures flourish in different regions of the world, making it a more colorful time. In the opinion of historians, five to seven thousand years ago there were approximately twenty different major civilizations, but over the course of time, many disappeared. Typical fascinating examples are the Incan and Mayan cultures of the Americas. They no longer play a role in modern society and are not likely to have an impact on the future of humanity. More recently, the ancient Australian aboriginal civilization has almost disappeared as a result of the impact of modern civilization. The ancient civilizations of Eurasia and North Africa did not suffer the same fate, instead continuously developing over

Facing Complex Systems: The End of Reductionism

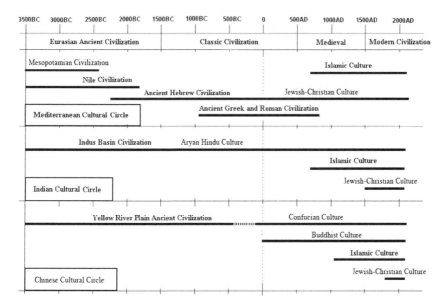

Figure 12.8. A brief history of the three major cultural circles in Eurasia.

the past two or three thousand years. They profoundly influence civilization today and will continue to do so in the future.

To streamline the discussion, with the assistance of figure 12.8, let us review the history of the development of the three major cultural circles in Eurasia. Relatively speaking, the Indian cultural circle is the simplest of the three because it has experienced the least change in its history. Conversely, the Mediterranean cultural circle, having changed a lot, is the most complicated. We will address these cultures in order of complexity.

Indian Cultural Circle

One of the reasons India's cultural history is the simplest is that it possesses the least and most unreliable historical data. While this is a seemingly a ridiculous reason for simplicity, it is a reality that must be faced. Although India has been subject to foreign aggression, occupation, and domination, first by the Indo-Iranians, then the Muslims, and finally the British, the influence of its relatively intact ancient culture is still readily apparent in everyday life and in the hearts and minds of the people.

Unlike China, which demolished its ancient culture several times, India's ancient culture was never subjected to the same self-destruction. This contributes to the Indian culture being the best preserved.

According to the literature, the earliest ancient culture of India can be traced back approximately 3,500 years to the Vedas, a large body of texts written in Sanskrit. This timeframe places them in the era of the Pentateuch, also known as the Torah or the first five books of the Old Testament. The Pentateuch is an almost historical text describing events occurring thousands of years ago and earlier. In contrast, the Vedas are a collection of poems that are not retrospective in nature. This leads many to believe that Indian culture is limited to 3,500 years, but it actually has a much longer history.

The original inhabitants of India did not write the Vedas; the invading Indo-Iranians were responsible. This is also true for the later tomes like the Bhagavad Gita, the Brahmanas, and the Upanishads. The origin of the Indo-Iranians is unclear. They might have come from Iran, Italy, or even Germany, but they were the first group of outsiders to invade India. That being said, the philosophy and worldview reflected in the ancient holy texts are completely different from cultures in Europe or those occupying the much nearer Middle East. Evidently, the Indo-Iranians in India absorbed a lot from the preceding local ancient Indus Valley culture.

The Indo-Iranians made a great contribution by recording their beautiful ancient culture, but as an ethnic minority ruling a local majority, they invented the extremely unpleasant caste system. This divided the population into four castes according to skin color, a practice even more distasteful than the Confucian hierarchy in China. While the caste system was formally abolished in 1947, the impact of its multiple millennia of existence still casts a shadow over present-day Indian society.

The most beautiful products of Indian culture include the principle of non-violence, the ideas of reincarnation and karma, and the theories of aura, chakra, and the seven-layer body. These ideas will continue to add a beautiful luster to world culture.

The Indian culture's concepts of aura, chakra, and the seven-layer body play a major role in this book and will have a far-reaching impact on the development of medicine and science. From this point of view, the purpose of this book is to appeal to more people, particularly young readers, to envision a better future for the development of medicine and science.

Chinese Cultural Circle

The Indian culture's beautiful notions of nonviolence, reincarnation, and karma were peacefully disseminated in China by Buddhist missionaries. They fused deeply with Chinese culture to the point where many present-day Chinese regard them as a part of Chinese culture. Given this, what is the origin of Chinese culture? Many would say Confucianism, but this is incorrect; Confucianism emerged only two thousand years ago.

Taoism is closer to the original Chinese culture. While it has only about 1,800 years of history as a religion, Taoism regards the Chinese philosopher Lao Tzu (571–470 BCE) as its original teacher. Lao Tzu was the contemporary of Confucius (551–479 BCE), who established the hierarchal philosophical base of Confucianism, which became the national religion of China in 134 BCE.

Lao Tzu was also the author of a famous text, the Tao Te Ching, which summarizes the philosophy of the Tao. The word *tao* in Chinese has two meanings: "speak" as a verb, and "the essence of the world" as a noun. Therefore, the word *tao* is similar to the Greek word *logos,* which has the same two meanings.

As the spirit of pursuing the essence of the world, the philosophy of the Tao is really the origin of Chinese culture. Long before the Tao Te Ching was the I Ching, as old as the Vedas and the Pentateuch. The I Ching laid the foundation for Taoist philosophy in China. Unfortunately, many years of civil war subsequently occurred—though not as a result of the Tao—and almost completely destroyed the peaceful Tao culture in China. Lao Tzu actually lived at the end of Tao period, and according to legend, he bemoaned the neglect of the Tao and left China, heading west.

In 221 BCE, China was finally unified by force. To assist in maintaining a highly centralized empire, Confucianism was established as the national religion by imperial decree in 134 BCE. Since that time, China has had a perfect unification of state and religion, far stronger than any that existed in Europe.

At that time in Europe, while the pope had theocratic authority, he had no military or political power. Conversely, numerous warlords and bandits enjoyed military power and political power but did not have theocratic legitimacy, and needed the pope to extend his authority to them by crowning them. This allowed them to become divinely appointed kings in the eyes of the population, and in exchange they collected the 10 percent tithe for the church.

In China, with the national religion of Confucianism in place, the emperor became the "sole son of heaven"—in addition to possessing political and military power, the emperor was also the theocratic authority. This era ushered in the Literary Inquisition, where intellectuals faced official persecution for their writing. This happened under each of the ruling dynasties and continues even today. Much like the Dark Ages in Europe, this discouraged thinkers and creative ideas, which impeded the development of science and technology.

It cannot be said that impeding critical thought and creative ideas under Confucianism served no purpose. The Literary Inquisition combined with a strict social hierarchy that made the central government very powerful. Europe and China were both unified for the first time in the second century BCE, but while Europe's unification was relatively short-lived, China remains unified to this day.

The second advantage of Confucianism's strict social controls is that it is very effective for military actions. Establishing Confucianism as the national religion solved the problem of harassment by nomads and greatly expanded China's territory. Unlike India, which was subjected to foreign rule for most of its history, China successfully assimilated outsiders into Confucianism on two occasions.

The success of Confucianism shared many similarities with the success of Islam, which also effectively used unification of thought to unify the Arab people, providing them strength and greatly expanding their territory. In addition to its great military victories, Islam was also notable for its success with internal management. The combination of unified thought, literary control, patriarchal social structure, and other factors allowed them to establish what some, depending on their definition of harmony, might consider a perfectly harmonious society. Naturally, this kind of military and internal success comes at the expense of freedom of thought, personal freedom, and even human dignity.

While Europe had the thousand-year-long Middle Ages of constrained thought and Islamic countries spent more than 1,400 years in a similar state, China has endured a similar state lasting over two thousand years. China is currently approaching the end of the medieval period, and thus a period similar to the Renaissance is just around the corner.

As Europe neared the end of the Middle Ages, its economy flourished, ushering in the abundantly rich and beautiful Renaissance. Among its other achievements, it bequeathed plentiful beautiful artworks to future generations and brought about the Reformation, which led to profound changes in thought.

One of these was the belief that everyone is a child of God. This change posed complications for governance: everyone might be a child of god, but not all of them could be kings. Ultimately, democracy became the system of governance for many nations.

Even though this period included much tension, anxiety, struggle, and violence, it was also a golden age that included the Renaissance in Italy and the Elizabethan era in England. Stubbornly entrenched outmoded ideas fighting to retain control slowly and incrementally made way for new ideas. The industrial revolution and the scientific revolution appeared in Britain and changed the world.

Given that China is at the end of its medieval period, there is reason to be optimistic that a wonderful new golden era is approaching.

Mediterranean Cultural Circle

Some Chinese scholars contend that the European and American culture, which dominates the world today, is a fusion of Hebrew and Greek cultures. Neither are original ancient cultures. As shown in figure 12.8, the Mesopotamian and Nile civilizations were established long before the Hebrew and Greek cultures.

Both the Tigris-Euphrates Valley, home to Mesopotamia, and the Nile Valley possessed geographic characteristics that left these civilizations overly vulnerable to nomadic invasion and destruction. As a result, both declined long before the classical era, known for its many great philosophers in both the East and the West. Travel and trade in the area led to many of the positive attributes of Mesopotamia and the Nile civilization being absorbed by the Hebrew and Greek cultures, which then fused and gradually evolved into the cultural base for Europe and the modern Americas.

Starting from Abraham, according to biblical history, Hebrew culture has existed for at most four thousand years. The original Hebrews migrated from the Tigris-Euphrates Valley to Canaan, located where modern-day Israel is. From a geographical viewpoint, it is evident that Israel occupied a channel or bridging position between the great Mesopotamian and Nile civilizations. Two hundred years later, the Hebrews migrated again, this time to Egypt, where they lived for four hundred years. When they ultimately returned to establish Israel, they had been deeply influenced by the Nile culture, which, combined with their Mesopotamian origins, gradually formed the unique Hebrew culture.

From the time of Moses (circa 1526–1446 BCE) to that of Malachi (circa 440–430 BCE), it took almost a thousand years for a unique and systematic Hebrew culture to form. Its nature is exhibited in the Hebrew Bible, the source of Christianity's Old Testament, which clearly and meticulously expresses their worldview and their faith. The dutiful, objective, and rigorous writing style of the Hebrew Bible also established a standard that future cultures could aspire to for recording their own history. Both the Hebrew Bible and the New Testament contain detailed records of many historical figures and events that can withstand strict archaeological examination.

In contrast, India's historical records are somewhat unreliable. Even where an important figure like Gautama Buddha is concerned, there is considerable uncertainty about when he was born and died. In different Indian literary sources, the discrepancy can be as large as a hundred years.

Even more notable and unique is the Hebrews' willingness to face their national scandals. For example, even when writing about their most celebrated hero, King David, they faithfully record all his faults. The Hebrew Bible vividly describes how he seduced another man's wife and then arranged for man's death. Similarly, almost all the prophets in the Hebrew Bible, whose writings constitute one-third of its length, are highly critical of their own people. That this level of cultural scrutiny and reporting of scandalous events is present in a holy text is particularly notable.

This acceptance of recording history that is critical of authority is not present in Chinese culture. Sima Qian (145–86 BCE), considered by many to be the greatest Chinese historian, suffered imprisonment and castration for recording work disputing government officials' rendition of historical events.

In addition to providing a standard for future historians and politicians to aspire to, the Hebrews' willingness to face the negative aspects of their history also inspired a worthy tradition in science. Nowadays referred to as the "spirit of science," its significance was eloquently summarized by the scientific philosopher Karl Popper (1902–1994):

> Science is one of the very few human activities—perhaps the only one—in which errors are systematically criticized and fairly often, in time, corrected. This is why we can say that, in science, we often learn from our mistakes, and why we can speak clearly and sensibly about making progress there.[1]

In 30 to 70 CE, Paul the Apostle brought Hebrew culture and the teachings of Christ from the Middle East to Europe. At that time, Greek had become the common language in the Roman Empire, occupying a status equivalent to the English language in modern times. Consequently, Paul and his Hebrew companions wrote the New Testament in Greek.

In a sense, the New Testament is a fusion of Hebrew culture and Greek culture, and it can be considered that first such fusion. Three centuries later, the New Testament and the Old Testament were put together as the Christian Bible, which served as the cultural basis for Europe and the United States and thus casts the largest influence on modern international culture.

The Greeks, also known as the Hellenes, exhibited both deep thinking and a practical attitude. Greece occupies mountainous terrain with very few plains, which impeded the development of a centralized empire and its inherent restraint on thought. It was formed by many small independent poleis, or city-states, that exhibited large diversity in thought. Historically, Greeks enjoyed significant freedom of thought, and they invented the world's oldest democratic system.

Greece's long Mediterranean coast facilitated travel and commerce, which, together with geographic proximity to the great ancient civilizations of Mesopotamia, the Nile, and Persia, allowed for significant cultural input. At the same time, it was sufficiently distant from these civilizations to be able to develop independently.

The decline of Mesopotamia, the Nile, and Persia coincided with the dawn of the classical era, where Greek culture rose to a position that matched the great periods of the ancient Chinese and Indian civilizations. The classical era produced many outstanding thinkers in the East, West, and South of the Eurasian continent. Given that the undeveloped state of travel and communication precluded any sort of international sharing of philosophical ideas, the reason for the simultaneous emergence of so many great thinkers poses an interesting question.

Underdeveloped travel and communication made it possible for different civilizations to develop different modes and frameworks of thinking. This is particularly true for cultures as geographically separated as those in the East and the West. Also, perhaps more importantly, the rise of ancient Greek culture illustrates that the coalescence of different cultures is often fruitful. The intersection of the Hebrews and the Greeks was one of these prosperous unions.

Over a thousand years later, European culture was influenced by a second fusion of Hebrew and Greek cultures. One of the major instigators of this subsequent interaction was the Crusades (1096–1291), a series of unsuccessful military campaigns with a deservedly tattered reputation. Perhaps their only positive historical contribution was that a large number of ancient Greek texts were brought back to Europe among the plunder. This outcome ultimately brought about the Renaissance.

In the early stages of the Roman Empire, Greek culture enjoyed a respected position, but after three hundred years, the Roman Catholic Church became the official religious authority of the Roman Empire and started its own version of the Chinese Literary Inquisition. This intolerance extended to existing Greek culture, which was viewed as heresy. Not wanting to be persecuted, many Greeks fled overseas, taking a large number of ancient texts with them. The subsequent centuries of upheaval saw the Europeans gradually forget the profound philosophy of Socrates, Plato, and Aristotle, along with the beautiful sculptures, paintings, poetry, and dramas as well as the science, medicine, and other riches of Greek civilization.

Encountering so many treasures of ancient Greek civilization had a profound impact on European society. In the context of the depressing life of Middle Ages Europe, ancient Greek culture was dazzling. This led to a movement to revive ancient Greek culture, which is the meaning of the word *Renaissance*.

At this time, Europe was in the later Middle Ages, and capitalism had already started to rise, accompanied by a new bourgeoisie. This burgeoning social class wanted more power and freedom, putting it at odds with the Vatican, which, after a thousand years of theocracy, had become very corrupt. Given an emerging class agitating for change and vested-interest groups trying to maintain the establishment, conflict was unavoidable. Starting with the publication of Dante Alighieri's *Divine Comedy* in 1321 and culminating in Britain's Glorious Revolution in 1688, the struggle lasted more than three hundred years. Aspects of it are referred to as the Renaissance and the Reformation. The outer layer of this major shift involved literary and artistic change, while the inner layer involved a deep change in the human mind.

The *Divine Comedy,* which represents the beginning of the Renaissance, is a long poem divided into three parts—Hell, Purgatory, and Paradise—that describe Dante's travels through these realms of the dead. Along the way he meets many departed souls, both virtuous and villainous, in their appropriate realms. It presented many inflammatory ideas, most notably consigning the contemporary pope and two of his recent predecessors as being condemned to hell.

The Reformation can be considered to stem from the efforts of John Wycliffe (1328–1384), a professor at Oxford University. He produced the first English version of the Bible and claimed that its authority was superior to that of the church. He also said that believers were accountable to Christ rather than the church. Obviously, these sentiments were not well received by the pope and church authorities. Thirty years after Wycliffe's death, he was declared a heretic, and it was decreed that his books be burned and his remains exhumed. In 1428, at the command of Pope Martin V, his remains were dug up, burned, and the ashes cast into the River Swift.

Another important character at the beginning of the Reformation was Jan Hus (1372–1415), president of Charles University in Prague. Influenced by Wycliffe, he promoted a similar view that later deeply influenced Martin Luther (1483–1546) and John Calvin (1509–1564). In 1414 Pope John XXIII investigated Hus and had the Archbishop of Constance imprison him, and eventually burn him at the stake, his ashes thrown into the Rhine River.

The Renaissance and Reformation contain too many similar grisly stories of persecution. Often described as a golden or great era, it was a turning point and a time of rapid change, where old ideas and systems staunchly resisted eviction from the historical stage. The birth of new ideas and systems exacted a high price in turmoil, conflict, and suffering.

Basic principles of the Renaissance and Reformation that spread incrementally from Europe included the idea that everyone is a child of God, with the associated right to dignity and authority. This inevitably led to the emergence of democratic systems and ideological emancipation. Freedom of thought allowed people to direct their intelligence where they wanted, and subsequently inventors and scientists emerged. In the United Kingdom, the first Protestant country, the industrial and scientific revolutions followed. In the end, these basic principles made European and American culture, with its roots in the fusion of Greek and Hebrew culture, the world's dominant culture.

The industrial revolution is not a theme of this book, but scientific revolution is. Modern science is the consequence of the Renaissance and Reformation, and thus it ultimately stems from the Hebrews and Greeks. The essence of modern science can be divided into the scientific spirit and the scientific method.

Put simply, the scientific spirit is the spirit of inquiry and getting down to the bedrock—essentially asking "why" endlessly followed by "why." In the words of Copernicus, the spirit of science is the spirit of pursuing the truth. It also encapsulates the spirit and courage of self-criticism and the resolution to criticize authority. As such, it can be considered a descendant of the Hebrew culture, which established the first model of this.

The scientific method incorporates the methods of severity, systematic approach, objectivity, rationality, logic, repeatability, quantification, numeration, and formulation. This series of methods was first established by Aristotle, perfected by Euclid in the *Elements,* and ultimately became the standard for modern science. Given its origins in ancient Greek culture, the scientific method can be considered the Greek method.

The success of modern science, technology, and industry as well as economic and military development has seen European and American culture dominate for the last five hundred years. Consequently, other countries with other cultures, regardless of their feelings about the dominant culture, have had to learn modern science and technology from Europe and the United States. They brought knowledge of science and technology home as an absolute truth. This led to a veneration of modern science to the effect that anything, regardless of its true merit, could become revered and beyond doubt as long as it was labeled scientific. Examples of this include so-called scientific socialism and the scientific development viewpoint.

The scientific method, based on ancient Greek culture, is easy to understand, handle, and express. It is also clear, diligent, and well organized. Conversely, the scientific spirit, based on ancient Hebrew culture, is difficult to express, teach, and learn. In terms of Chinese philosophy, it can be thought of as Tao instead of skill. In the words of Lao Tzu, "The Tao that can be told is not the eternal Tao."[2] Thus, the great majority of those who traveled to Europe and the United States to study science and technology learned only the Scientific Method without picking up the scientific spirit.

Even worse, those who returned having learned only the scientific method brought it back to enshrine as a pillar of unquestionable truth. This led to beautiful, vivid, and dynamic modern science with its spirit of exploration and self-criticism being transformed into an ugly and stern relic, existing to be worshipped and to intimidate those who dare question its power and reliability.

Facing Complex Systems: The End of Reductionism

The deification of modern science is evident when the theoretical framework of Classical Chinese Medicine conflicts with the framework of modern science. Typically the questions raised are along the lines of, "Is Chinese medicine scientific or not?" and "Is it possible to prove Chinese medicine by means of modern science?" Questions that are seldom heard include, "What new issues can Classical Chinese Medicine raise in modern science to enrich it?" and "What challenges can Classical Chinese Medicine deliver to modern science in order to further develop it?" Even rarer are, "What deficiencies in modern science are revealed by the practice of Chinese medicine?" and "How can the research of Classical Chinese Medicine further develop modern science?" These last four questions are the main themes of this book.

In the context of culture, the underpinnings of modern science can be considered the Hebrew spirit and Greek methodology. It contains neither ancient Chinese nor Indian cultures, both of which are reputable and possess great wisdom. Therefore, the important issue is whether we can integrate the Hebrew spirit with ancient Chinese methodology and Indian methodology to further promote the development of modern science. From the perspective of medical science, the different aspects of the cultures can be described as follows: Hebrew culture provides the soul, Greek culture provides a body with flesh and blood, Chinese culture provides the breath of life, and Indian culture provides the beautiful aura of life.

In literature it is impressive when a writer can vividly describe the appearance of a person. However, if we carefully consider the person described, it is a body with flesh and blood but without breath, aura, and soul. It is not a living person, but a walking corpse. This is exactly how the human body is described in Western medicine and modern science. It is also how it is described in ancient Greek culture. If the Hebrew, Chinese, Indian, and Greek cultures can be combined, it might be possible to develop a more complete understanding of ourselves as well as a truer representation of the world.

This book, especially the final three chapters, represents the outcome of the collision and fusion of various cultures. The result is to scientifically, holistically, and quantitatively measure and calculate the concepts of beauty and coherence in order to enrich a dawning new golden age. The intended purpose of this book is to introduce some existing relevant knowledge of physics to enable the ancient Chinese and Indian medical systems to be seen from the perspective of modern science. It might be more important to hope that readers can observe that when ancient Chinese and ancient Indian cultures collide with modern science, beautiful new growth can arise.

I believe we are at the dawn of a new golden age. It is my hope that this book serves to provide a paving brick in a new path that leads to a new, more expansive, and more beautiful realm of science—a new heaven and a new earth.

CHAPTER 13

HOW MUCH BEAUTY IS THERE IN A BALLET?

Pure mathematics is, in its way, the poetry of logical ideas. —ALBERT EINSTEIN

Professor Zhong-shen Liu, a close friend of mine, is a pharmaceutical expert at Heilongjiang University of Traditional Chinese Medicine in Harbin, China. In 1986 he made the long trip to Hangzhou, my hometown, for a conference. I took the opportunity to invite him to participate in an interdisciplinary discussion group at Zhejiang University and to present a talk to the group.

Like many of his colleagues, his research involved performing chemical analyses of herbal medicines. This usually involves using liquid and gas chromatography to separate an herbal remedy into its thousands of constituent chemical components, then testing each of them individually to determine which one is the key component. Once the decisive component has been identified, the next step is to determine the chemical structure of the relevant molecule in order to develop a way of synthesizing it in laboratories, and then economically in factories.

These steps represent the typical successful approach to studying herbal medicines at that time and still apply today in China and many other countries. The process is consistent with reductionism, which has dominated biology and medicine for hundreds of years, particularly in the twentieth century.

Natural Medicine Is Music instead of Mechanics

Liu shared his personal research experience with us. In some cases the reductionist approach was successful in finding the decisive component, but in other cases they could not find the decisive component at all. This shows that the potency of some herbal medications lies in the combined effects of many components rather than a single one.

His opinion resonated strongly with the members of our group, particularly since we had just been discussing the problems of reductionism in biology and medicine. A young student of the arts, Jin Fu, even likened this approach in studying Chinese herbal medicine to trying to study music by means of reductionism—trying to determine which note is most important and decisive in the music of Mozart, and considering how to discard all the other less important notes. Thus one could attempt to compose a purified symphony with only a single note, even with only a single frequency, which, in this method, would be the essence of Mozart's music.

While this satirical idea is extreme, it does reflect the essential nature of herbal medicine research and conventional medicine in general. Millions of scientists spend their whole lives looking for the decisive factors of herbal remedies or the singular causes of diseases, like tirelessly looking for the decisive note in a piece of music.

The reductionist approach is not completely misguided, and has proved extremely successful in biological and medical research. Almost every infectious disease is caused by a unique species of bacteria or virus, and Western medicine has been successful in conquering epidemic deceases. However, medicine has changed, and acute bacterial diseases are relatively rare. Instead, people, particularly those in developed countries with good medical systems, suffer from chronic diseases and functional disorders. Many have not found relief in Western medicine and have looked to alternative therapies like homeopathy, acupuncture, ayurveda, massage, Qigong, and other natural forms of medicine.

The underlying principle of many natural forms of medicine is holism rather than reductionism, and the key point of holism is coherence. The word *coherence* is used in music and other arts but rarely in science because it is difficult to comprehend from a rational, quantitative perspective. Fortunately, studying coherence is not as hopeless as generally believed, as shown in the following discussion.

Two Levels of Problems in a Symphony

The first precondition for performing a successful symphony is that every musical instrument is well made and in good condition. A single faulty instrument immediately damages the coherence of the symphony. In such a case, to fix our orchestra we must discern which instrument is out of order and repair it. At this level, which we can refer to as single-factor thinking, reductionism works perfectly.

The second precondition is that all the musicians cooperate and coordinate well. Even if every instrument is in good condition and each musician plays the piece properly, a lack of coordination will ruin the symphony. At this level, single-factor thinking and reductionism no longer work.

It is evident that Western medicine has focused on the first precondition for creating a beautiful symphony. The modern challenge for medicine is to move toward being able to fix the symphony when the second precondition is not being met, when coordination and cooperation are not functioning properly. This poses a second-level challenge, a major challenge to the human intellect. It is a much more difficult task for scientists to objectively and quantitatively analyze something as complicated as the body-mind system.

The Role of Order in Music

As the challenge facing medicine shares some common characteristics with music, let us consider whether it is possible to scientifically and quantitatively evaluate the degree of coherence in music and dancing. The first step is to contemplate order.

In the 1970s, the concept of order became fashionable in both science and art. Interestingly, the antonym of order, chaos, has also become fashionable. Musicians believed that there is some implicit order in music that makes it pleasing to the ear. Therefore, the higher degree of order presented, the better the music is. Similarly, medical doctors believed that living systems are always highly ordered, and that a higher degree of order denotes a healthier individual. These beliefs are not far from the truth, but when we start to consider coherence, the analysis becomes more complicated.

The 1990s saw more and more scientists studying the concept of coherence or the coherent state. Since scientists' understanding of coherence is constantly evolving, it is difficult to categorically define exactly what coherence is. However, the following three examples of idealized states of coherence illustrate the coherent state.

The first example is the perfectly chaotic state, such as a group of very young children in a nursery without adult supervision. Too young to participate in cooperative play, there is no coordination in their actions. This state of coherence represents an almost perfectly chaotic state. In such a state, one hundred children have one hundred degrees of freedom, meaning one hundred parameters that can vary independently.

Soldiers in an honor guard represent an example of the opposite of a chaotic state, exhibiting the highest degree of order, all acting exactly the same, like a single person. This state is a perfectly ordered or "crystal" state, and the one hundred soldiers have only one degree of freedom.

The third state is approximated by dancers in a ballet. They are not acting chaotically, like children in a nursery, nor are they perfectly ordered, like soldiers in an honor guard. Their cooperation is both dynamic and coherent. Participants are free to move independently, but their actions are perfectly coordinated. Scientists view this as a coherent state, neither highly ordered nor chaotic.

Coherence: Neither Highly Ordered nor Chaotic

Music represents the combination of many acoustic frequencies. In addition to being dissected into a combination of many individual notes, music can also be dissected into a combination of many frequencies (fig. 13.1). When frequencies are combined, the fundamental frequency designates the note. For example, the fundamental frequency of C is 532 hertz, while A has a fundamental frequency of 440 hertz. As discussed in chapter 8, the fundamental frequency is complemented by many higher frequency overtones (fig. 8.2). The frequencies of the overtones are usually a whole number of times larger than the fundamental frequency, so A, with its fundamental frequency of 440 hertz, has a first overtone that is twice the fundamental frequency, or 880 hertz, a second overtone three times the fundamental frequency, or 1,320 hertz, and so on. However, the combination of the strengths of overtone frequencies varies by instrument and from person to person. This is what gives each person and each instrument its distinctive sound, even when they are playing the same note.

Music in a state of highest order exhibits only one frequency, the fundamental frequency. In a chorus, the highest order occurs when all singers sing the same note and frequency, similar to the movement of soldiers in an honor guard. This does not represent coherent music, however. Few would want to listen to a completely random combination of frequencies that would typify a perfectly chaotic state of sound, and it would be difficult to classify it as music; most would call it noise. Science goes one step farther, classifying it as ideal noise, or if the strengths of all the frequencies are similar, white noise. In coherent music, the relationship between the different notes and frequencies is in a

How Much Beauty Is There in a Ballet?

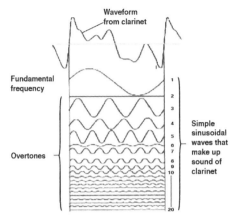

Figure 13.1. Frequency analysis of one note of music.

Figure 13.2. How many dancers are in the picture?

coherent state, similar to the relationship among dancers in a ballet. Thus the question becomes, what is it about the combined actions of dancers in a ballet that creates coherence?

1 + 1 = 3: The Miracle in Coherence

The task of calculating the degree of coherence is not as daunting, and the mathematics involved not as painful, as one would suppose. The fundamental principle behind the calculation can be initiated with a simple count. The following example illustrates the concept. Let us start with the question, "How many dancers are there in figure 13.2?"

Figure 13.3. A duel can result in 1 + 1 = 0.

On a basic level, it is a straightforward question, and short of some disguised dancers lurking in the background, the answer, two, is evident. If, however, instead of thinking in terms of the number of humans on stage and instead consider the number of possible forms, the answer is more variable. For instance, when the two are dancing in perfect cooperation, it is sometimes hard for us to tell which arm and leg belong to whom. It begins to appear that a new creature, a third dancer with four arms and four legs, emerges from such perfect cooperation. The most appropriate mathematical formula to describe the situation is $1 + 1 = 3$, instead of the ordinary $1 + 1 = 2$.

In fact, there are at least four possibilities or four different answers to the most basic mathematical formula $1 + 1$. For instance, a duel could result in 1+1=0, if both parties fire at the same moment and hit one another. $1 + 1 = 0$ can also be observed in wars, in business battles, and even within families.

Of course, most try to avoid an outcome of $1 + 1 = 0$. In addition to this and to $1 + 1 = 2$, there is also the situation of $1 + 1 = 1$, which is also quite common in history, as when a victor successfully occupies and eventually annexes a losing country. Similarly, in business, when a company outcompetes and eventually merges with a rival, $1 + 1 = 1$ applies, providing the companies are lucky enough to avoid the $1 + 1 = 0$ scenario.

The question then becomes, is there a situation that is not only better than $1 + 1 = 0$, but also better than $1 + 1 = 1$ and $1 + 1 = 2$? There is: $1 + 1 = 3$ is also possible. So three different answers to $1 + 1$ are possible, depending on the type of mathematics used.

How Much Beauty Is There in a Ballet?

Figure 13.4. Children in a nursery: 1 + 1 = 2

Three Ideal States

The formula $1 + 1 = 0$ is not relevant to this discussion, as it describes the state of dying. The three different formulas that can be used to describe the three idealized states in living systems are:

For a chaotic state, like children in a nursery without supervision: $1 + 1 = 2$

For a crystal state, like soldiers in an honor guard: $1 + 1 = 1$

For a coherent state, like ballet dancers: $1 + 1 = 3$

Let us discuss the three ideal situations individually.

Ideal Chaotic State: 1 + 1 = 2

In this state, the participants act completely independently, like the behavior of young children in a nursery without any supervision in a perfect chaotic state. The basic mathematical formula is the everyday $1 + 1 = 2$, and the extensions for situations with more participants is also straightforward:

For two children: $1 + 1 = 2$

For three children: $1 + 1 + 1 = 3$

For four children: $1 + 1 + 1 + 1 = 4$, and so on.

In terms of physics, in a perfect chaotic state, one hundred children have one hundred degrees of freedom.

Figure 13.5. Soldiers in an honor guard: 1 + 1 = 1.

Ideal Crystal State: 1 + 1 = 1

The second state is like soldiers in an honor guard (fig. 13.5), in the highest order and acting exactly the same, like a single person. The basic mathematical formula is also simple: 1 + 1 = 1. Extension of 1 + 1 = 1 is also very simple:

For two soldiers: 1 + 1 = 1

For three soldiers: 1 + 1 + 1 = 1

For four soldiers: 1 + 1 + 1 + 1 = 1, and so on.

The situation is as simple as it appears. In terms of physics, in a perfect ordered state, one hundred soldiers have only one degree of freedom. While it sounds counterintuitive that one hundred individuals could collectively only have one degree of freedom, this approximates the goal of dictators, and many of them come close to achieving it. It is also undeniable that this crystal state works very well for a war, even though it is predicated on sacrificing human freedom and human rights.

Ideal Coherent State: 1 + 1 = 3

The third state is like dancers in a ballet (fig. 13.2). They are neither as chaotic as the children in the nursery, nor as highly ordered as the soldiers in the honor guard. Each dancer moves independently, but at the same time coordinates their actions. In the case of two dancers, in addition to their individual forms, their cooperation produces a third form, which can be thought of as another dancer. Scientists call it a coherent state: it is neither highly ordered nor chaotic.

How Much Beauty Is There in a Ballet?

Figure 13.6. Dancers in a ballet: 1 + 1 = 3.

Figure 13.7. How many dancers are in these pictures?

How Many Degrees of Freedom in a Coherent State?

How Many Dancers in a Ballet?

The basic idea behind the calculation can start with a simple count. In figure 13.6, the calculation is a little complicated, but not too difficult if you start with the simplest case, and then incrementally increase the complexity. The mathematics involved is called combinatorics. Two dancers acting coherently produce a third form, thereby allowing three degrees of freedom. This outcome can be calculated using the following general equation:

For n number of dancers, the number of possible combinations = $2^n - 1$

Thus in the case of two dancers, the number of possible combinations = $2^2 - 1$

As shown in figure 13.8, if three dancers perform together with excellent cooperation, there are seven possible combinations, or $1 + 1 + 1 = 7$. If we apply

179

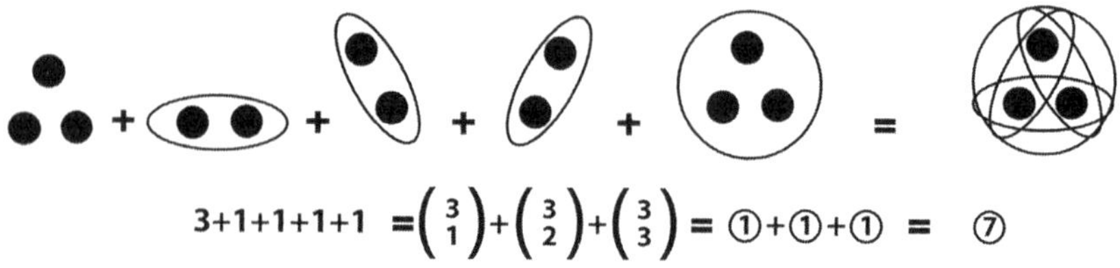

$$3+1+1+1+1 = \binom{3}{1}+\binom{3}{2}+\binom{3}{3} = ①+①+① = ⑦$$

Figure 13.8. The number of combinations with three dancers.

the general equation, for three dancers, the result is 7, from $2^3 - 1 = 7$. Similarly, for four dancers, $1 + 1 + 1 + 1 = 2^4 - 1 = 15$, and so on. By applying the formula it is relatively straightforward to calculate the extensions of 1+1=3.

Extensions of 1 + 1 = 3

For two dancers: $1 + 1 = 2^2 - 1 = 3$

For three dancers: $1 + 1 + 1 = 2^3 - 1 = 7$

For four dancers: $1 + 1 + 1 + 1 = 2^4 - 1 = 15$

For five dancers: $1 + 1 + 1 + 1 + 1 = 2^5 - 1 = 31$

For six dancers: $1 + 1 + 1 + 1 + 1 + 1 = 2^6 - 1 = 63$

For seven dancers: $1 + 1 + 1 + 1 + 1 + 1 + 1 = 2^7 - 1 = 127$

So what about one hundred dancers? The extraordinarily large number is:

$2^{100} - 1 = 126{,}750{,}600{,}228{,}229{,}401{,}703{,}205{,}375$

This means that instead of having just one hundred degrees of freedom, as if they were operating completely independently, in a perfectly coherent state they possess over 126 billion quadrillion different combinations or degrees of freedom. This is the miracle of coherence.

In 1991, just five years after my discussions with Liu about music in medicine, my team in Germany developed the key to analyzing coherence in music and in living systems. The first academic paper presenting rigorous mathematical proof of this idea was published in the United Kingdom in 1994 in the journal *Medical Hypotheses*, and then republished in Switzerland in a special issue of the *International Journal of Modeling, Identification and Control* on physiotherapy.[1]

The principles and discoveries discussed in this chapter were applied to develop an instrument, called a coherence meter, which measures the state of the electromagnetic body in individuals. The following chapter looks at how this mathematical thinking can be practically applied to medicine and health care.

CHAPTER 14

MEASURING THE INVISIBLE RAINBOW

> Suppose that we are asked to arrange the following into two categories: distance, mass, electric force, entropy, beauty, melody. I think there are the strongest grounds for placing the entropy alongside beauty and melody.... Entropy is only found when the parts are viewed in association, and it is by viewing or hearing the parts in association that beauty and melody are associated.... It is a pregnant thought that one of these association should be able to figure as a commonplace quantity of sciences.
>
> —ARTHUR STANLEY EDDINGTON, *The Nature of the Physical World*

The British astronomer Arthur Eddington is credited with several important discoveries, including the relationship between the mass and luminosity of stars. He was an early advocate and popularizer for the general theory of relativity, a contentious hypothesis at the time, and provided the first experimental results that supported its predictions.

Eddington's profound insight, presented at the start of this chapter, reveal an early scientific consideration of a shared quantity related to entropy, beauty, and melody. More specifically, he discusses a shared quantity related to the association of parts. The calculation of coherence, outlined in the preceding chapter, relates to this idea.

Modern science has naturally developed from the study of simple systems to the study of more complex systems. The solar system, thoroughly studied and described by Copernicus, Galileo, Kepler, Newton, and

Figure 14.1. Arthur Stanley Eddington (1882–1944).

many others, is a simple system. Similarly, an atom, which in many ways is like a miniature solar system, is also simple in structure. The human body is a considerably more complicated system.

To summarize the earlier discussion, traditional reductionist approaches to studying such a complex system involved portioning it into progressively simpler systems. While highly successful in meeting crucial medical challenges of the past, the progressively closer focus on individual components came at the expense of our understanding of the interactions among them. In the face of the chronic degenerative diseases stemming from functional disorders that currently challenge health care systems, the reductionist approach is of limited use. Consequently, holistic medical systems have enjoyed a dramatic increase in demand in recent times.

In order to move forward with treating and preventing chronic degenerative conditions, we have to refocus our research on the interactions of the components of living systems. Analyzing coherence is key to progressing this research. While quantitatively assessing and evaluating the degree of coherence once posed a daunting, some believed insurmountable, challenge, scientists have devised methods to achieve this goal.

The preceding chapter introduced coherence theoretically and mathematically. The next step is practically measuring the coherence of a body-mind system. The process of discovering the invisible dissipative structure of electromagnetic fields in living systems provides an insight into how analysis of coherence can be performed. This analysis not only enables insight into a new aspect of the human body but also provides a practical new method to instantly evaluate the condition of a person, holistically and quantitatively. This is achieved by measuring the energy distribution of the invisible electromagnetic structure in individuals.

Coupling Oscillators and Energy Movement

The dissipative structure of electromagnetic fields in living systems is composed of chaotic standing waves. To explore how coherence operates in this situation, let us first investigate a simplified example of how the degree of coherence can be measured from the study of the coupling of waves. Coupling refers to the degree to which two oscillating bodies are constrained to move cooperatively.

In terms of coherence, the lungs, considered as a single organ, and the heart can be regarded as something akin to two dancers, because they are perma-

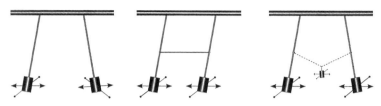

Figure 14.2. Coupling relationships of two pendulums.

nently pulsating with their independent rhythms but still require a degree of coordination. This will enable their degree of coherence to be evaluated.

Given their differing functions, a situation where they were acting identically, in a highly ordered state, would obviously be detrimental to the living system. Similarly, working completely independently, as in the chaotic state, would also be far from ideal. For example, when someone is running, both the heart and the lungs need to work harder than usual, even though they do so at different frequencies. This means that there must be some coupling between the heart and the lungs that is neither too strong nor too weak. Flexible coupling enables coherence.

From the viewpoint of physics, the heart and the lungs can be modelled as a pair of oscillators, for example as a pair of pendulums. Figure 14.2 depicts three important coupling relationships that can exist between two pendulums. The first relationship, at left, shows no coupling, which can be referred to as zero coupling. The two pendulums move completely independently. Their oscillations emit two independent waves with differing frequencies. This represents an ideal chaotic state.

In the middle image, the two pendulums are connected by an inflexible connector. This results in very strong coupling between them. The two pendulums oscillate as a single body and emit waves of only one frequency. This represents the ideal crystal state, like the soldiers marching in an honor guard.

The image on the right depicts flexible coupling, provided by an elastic cord with a weight suspended from it. The two pendulums are neither rigidly connected not completely independent, and are in a coherent state.

In figure 14.3, y_1 and y_2 show examples of two waves of different frequencies that could be emitted by each of the pendulums oscillating either independently or with flexible coupling. If the two pendulums are flexibly coupled, constructive and destructive interference leads to the development of a third frequency,

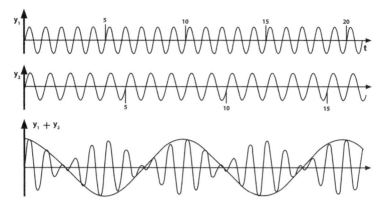

Figure 14.3. A beat frequency occurs when two pendulums are flexibly coupled.

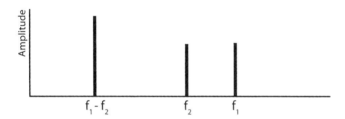

Figure 14.4. Amplitude of the frequencies.

labeled $y_1 + y_2$. This is referred to in physics as the "beat" frequency. As with the two dancers in a ballet, the unusual formula $1 + 1 = 3$ applies.

Figure 14.4 expresses the frequencies of the waves in figure 14.3 in a frequency spectrum. The beat frequency $(f_1 - f_2)$ is in fact equal to the difference between frequencies 1 and 2. The vertical axis shows each of the waves' amplitude, which is proportional to the energy level. In sound waves, higher amplitude means a louder sound. The higher amplitude of the beat frequency shows that flexible coupling has transferred energy from the higher original frequencies to the lower beat frequency. This is one of the important phenomena in coherence.

So far, we have considered only a limited case of coherence, between two oscillators or waves. When an infinite number of oscillators emitting an infinite number of waves is considered, the frequency spectrum no longer shows separate discrete frequencies like those in figure 14.4. It instead shows a continuous frequency distribution (fig. 14.5), which is a combination of countless waves.

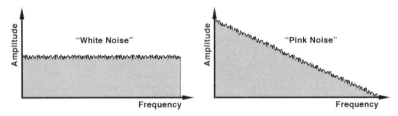

Figure 14.5. Frequency spectra of white noise (left) and pink noise (right).

Figure 14.5 shows the spectra of two typical combinations of a nearly infinite number of waves. On the left is the outcome of acoustic waves combining in an ideal chaotic state. The amplitudes of all frequencies are almost the same. In technology, this ideal chaotic combination of acoustic waves is called white noise. On the right are acoustic waves combining in an ideal coherent state. In technology, this combination was called "1/f noise" but was later also commonly referred to as pink noise. In comparison to white noise, the energy in pink noise has moved from the higher frequencies to the lower frequencies. This is consistent with the simple two-oscillator example in figure 14.3. This leads to the conclusion that in pink noise, there are many flexible couplings among the oscillators emitting these acoustic waves.

Pink noise was first detected emanating from an electronic circuit in 1926. It took physicists more than fifty years to determine that the underlying mechanism emerges from the grouping movement of electrons in a circuit. Interestingly, scientists working with music found that the frequency distribution of most classical and folk music was consistent with pink noise, while the frequency distributions of some modern forms of music are similar to white noise.

Modern technology makes it straightforward to analyze the frequency spectrum of acoustic waves; all that is required is a digital recording device and a computer to apply a mathematical technique known as the Fourier transformation, invented by the French mathematician Joseph Fourier (1768–1830). However, studying the frequency spectrum of electromagnetic waves in human bodies and in other living systems is significantly more challenging. The frequency range of electromagnetic waves in living systems is very broad, ranging from extremely low-frequency (ELF), on the order of 0.5 hertz, to the ultraviolet, at 10^{17} hertz. In comparison, audible acoustic waves occur in the relatively limited range of 20 to 20,000 hertz. This range is within the recording ability of acoustic sam-

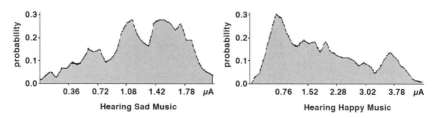

Figure 14.6. Energy distribution changes when the test subject listens to sad music (left) and happy music (right).

plers and the processing capacity of modern computers. Conversely, the range of electromagnetic frequencies present in living systems, from 0.5 to 10^{17} hertz, is so broad as to be virtually infinite. As such, it is beyond the ability of present technology to comprehensively detect or process.

Fortunately, physicists have found that skin conductivity, which is relatively easy to measure on the surface of the body, is proportional to the strength of the electric field. Thus measuring skin conductivity allows us to observe the distribution of electromagnetic fields in the body. This is determined by the dissipative structure of electromagnetic fields generated by the superposition of chaotic standing waves inside the body.

In other words, it is not necessary to measure the individual frequencies of electromagnetic waves to produce a frequency distribution, as is done for acoustic waves. Instead, the cumulative result of the superposition of the countless, even infinite, electromagnetic waves within a living system can be measured and analyzed.

Through history, mathematicians have developed statistical methods for studying the final result of the combination of infinite factors. These methods only require a finite test sample and around one hundred sample measurements to obtain meaningful results. The method, called probability distribution, makes it possible to observe changes in the energy distribution within the human body. Probability distributions generated from body conductivity measurements, depicted in figure 14.6, show that the dissipative structure of electromagnetic fields inside a body is very sensitive to changes in the music being listened to. In fact, it is quite sensitive to many psychological and physiological disturbances.

To understand the implications of these results, a couple of questions have to be addressed. First, what is the meaning of a probability distribution? Second, what is the relationship between probability distribution and coherence?

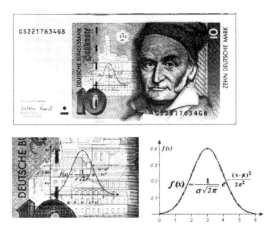

Figure 14.7. Carl Friedrich Gauss and Gaussian distribution portrayed on a German banknote.

Idealized Energy Distribution

Interestingly, long before the scientific study of coherence, mathematicians developed methods to describe its three ideal states—the chaotic state, the crystal state, and the coherent state.

1 + 1 = 2: Gaussian Distribution in an Ideal Chaotic System

Prior to the introduction of the euro, the old ten-mark German note (fig. 14.7) bore a portrait of Carl Friedrich Gauss (1777–1855). He was born into a poor family but was noticed as a prodigious talent in his early years; a wealthy aristocrat financed his education. Gauss started making profound contributions to mathematics from his student days and became an influential and famous professor at the University of Göttingen. He is widely regarded as one of the greatest mathematicians of all time.

The banknote also depicts one of Gauss's greatest contributions to mathematics, shown as a small curve with a mathematical formula (lower left in fig. 14.7) adjacent to his portrait. The curve portrays Gaussian distribution (fig. 14.8), a symmetric curve that is widely used for statistics in scientific research, industry, and other areas. It is such a standard everyday tool in statistics that it is referred to as "normal distribution."

It was not until recently that people became aware that Gaussian distribution also describes a perfectly chaotic state. The basic hypothesis behind Gaussian

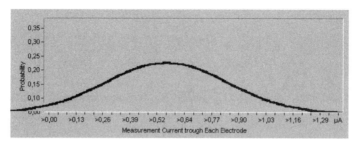

Figure 14.8. Gaussian distribution, from a system in the ideal chaotic state.

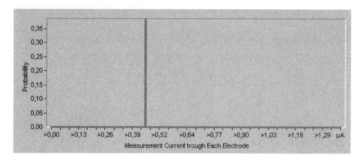

Figure 14.9. Delta distribution, from a system in the ideal crystal state.

distribution is that measurement data is influenced by an infinite number of factors, and that all of these factors are independent of one another. As such, the underlying hypothesis describes the state of those children in a nursery without supervision described in chapter 13. In other words, Gaussian distribution describes an ideal chaotic system.

Practically, this means that if a collection of measurement data conforms to a Gaussian distribution, as in figure 14.8, the data comes from a system composed of infinite independent elements, and is thus an ideal chaotic system. While no real system exists in a perfectly ideal chaotic state, Gaussian distribution is approximately applicable to many nonliving systems and is widely applied in this context.

1 + 1 = 1: Delta Distribution in an Ideal Crystal System

In addition to Gaussian distribution, there are various other probability distributions in mathematics, and one that is relevant to this discussion is delta distribution, perhaps the simplest probability distribution in mathematics. It resembles a narrow pillar or an upside-down T (fig. 14.9), which means that all the measurements are identical.

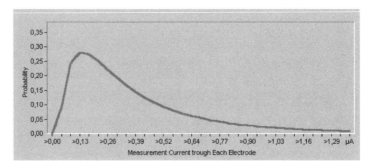

Figure 14.10. Log-normal distribution, from a system in an ideal coherent state.

If the data conforms to a delta distribution, all the system's elements act with the highest degree of order—all acting as one. A system described by a delta distribution is like the soldiers in the honor guard, exhibiting a crystal state. Clearly no living system would function like this, but this probability distribution provides a point of comparison for the real state of an individual.

1 + 1 = 3: Log-Normal Distribution in an Ideal Coherent System

In terms of living systems and the concept of coherence, the most important probability distribution of measurement data is the log-normal distribution (fig. 14.10). Log-normal distribution presents a nonsymmetric curve. Compared to the symmetrical Gaussian distribution shown in figure 14.8, the peak of the curve is shifted to the left. Log-normal distribution was considered unimportant until 1969, when the German mathematician Lothar Sachs found that the measurement data from many physiological systems was inconsistent with Gaussian distribution and instead conformed to log-normal distribution. This revealed that there is some relationship between log-normal distribution and living systems.

In 1994 I proved mathematically that log-normal distribution comes from an infinite-elements system in which each element has the ability to function independently, and at the same time, is also able to cooperate with other elements. In other words, if the measurement data from a system conforms to log-normal distribution, this demonstrates coherence among the elements. This means that the unusual arithmetic that describes a ballet, 1 + 1 = 3, applies. Consequently, the beauty and coherence present in music, ballet, and living systems is no longer limited to eloquent poetic description; it now exists in the realm of rigorous mathematical formulas, practical measurement, and quantitative calculation.

Infinite Dimensional Space and the Coherence Pyramid

This book has looked at three ideal states: the ideal chaotic state, like children in a nursery; the ideal crystal state, like soldiers in an honor guard; the ideal coherent state, like dancers in a ballet. We have also discussed the three types of probability distributions that correspond to these ideal states: Gaussian distribution, delta distribution, and log-normal distribution, respectively.

All three states represent ideal states. In fact, no real system is so chaotic as to be perfectly described by Gaussian distribution, nor so highly ordered as to perfectly conform to delta distribution, nor so perfectly coherent that it is perfectly consistently with log-normal distribution. A real system always exist somewhere between these three. Consequently, the probability distribution of its measured data is also somewhere between the three typical probability distributions.

The frequency distribution on the right in figure 14.6 shows a real distribution of data that is quite close to log-normal distribution, but not exactly the same. This means that the person being tested was in a very coherent state when listening to happy music, but not a perfectly coherent one. This raises the practical challenge of quantifying how far the state of a person is from a perfectly coherent one. While quantifying distance from coherence seems counterintuitive, much like doing so for beauty would be, this is the challenge at hand.

The techniques required to tackle the contemporary challenge of determining a coherence distance have again been provided by a great German mathematician, David Hilbert (1862–1943), shown in figure 14.11. He developed the mathematical tools to calculate the distance between two states that are composed of infinite elements.

Like Gauss, Hilbert was also a professor at the University of Göttingen, and in many ways he can be considered Gauss's successor as one of the most influential mathematicians of the era. At the beginning of the twentieth century, he presented a talk at a conference about the development of mathematics in which he predicted the main focus for mathematicians. This talk ultimately served as a navigation chart for mathematicians worldwide in exploring the unknown areas of mathematics.

Figure 14.11. David Hilbert, who theorized infinite dimensional space.

Hilbert's most important discovery is the concept of infinite dimensional space, which mathematicians call Hilbert space or space of functions. In Hilbert space, there is generalized distance that makes it possible to measure the "distance" between two mathematical functions that are both composed of infinite elements.

In everyday language, distance refers to a measurement between two points, like on the surface of a piece of paper or across a room. Hilbert generalized the concept of distance from real-world two- and three-dimensional spaces to an imagined four-dimensional space, five-dimensional space, and so on, extending it to n-dimensional space, which could even be infinite dimensional space. The calculation of distance in all these spaces is the same.

This infinite dimensional space allows for the distance between the real state of a patient and the ideal coherent state to be calculated. In other words, with Hilbert's help, it is possible to quantitatively state how far away a patient is from a state of perfect coherence. In addition to calculating the distance, it is also necessary to have some kind of coordinate system to stipulate the precise position of the state of a patient relative to the three ideal states.

The three ideal states are described by three rigorous mathematical formulas and a practical method of measurement. Hilbert space allows a mathematical formula or function to be condensed into a mathematical point. By condensing the three probability distributions, and thereby the three ideal states, into three mathematical points, a special coordinate system can be established (plate 16 in the color plate section). This allows the location of the real state of a person to be perceived.

As shown in plate 16, the special coordinate system in Hilbert space that allows us to evaluate the degree of coherence resembles a pyramid. Consequently it is referred to as a coherence pyramid. At the top of the pyramid is a green dot, the coordinate of the ideal coherent state as expressed by log-normal distribution. A black dot, marking the ideal chaotic state as expressed by Gaussian distribution, is at the left corner. The right corner of the coherence pyramid has a red dot, marking the ideal crystal state expressed by delta distribution.

To explain the choice of colors, black is the color of anarchy. Physiologically and psychologically, anarchy possesses both positive and negative aspects. The positive aspect is that anarchy provides relaxation to the system after hard work or conflict, and as such it is beneficial to health. On the negative side, anarchy sometimes steers a system into chaos or depression, the accumulation of which can be dangerous to the body-mind system. It can be argued that many cancers stem from the accumulation of chaos and depression.

Red is the color of revolution or dictatorship. Physiologically and psychologically, dictatorship also has two sides. On the positive side, it brings a system into very high order, which is important for conflict or for highly focused work. However, this focus comes at the expense of increased stress. Accumulated stress can lead directly to hypertension. It can also lead to a breakdown in the system, leading to chaos and its associated health risks.

Green is generally associated with ecological balance or a positive state, and has therefore been applied to denote an idealized state without negative implications.

CHAPTER 15

COHERENCE IN MEDICINE AND HEALTH CARE

> One should be aware of the constant presence, at least in Western philosophy, of a line of thought for which form, structure, relation, and the like were more basic than matter, energy, or substance. This line of thought, which goes back to the Pythagorean, was mostly succumbing in front of the bold partisans of matter and energy, especially when matter and energy proved to be exploitable for improving man's standard of life (besides having other less desirable but equally conspicuous uses). —GIUSEPPI LONGO, *Information Theory*

The East and the West both possess a history of holistic thinking. In the West, philosophy concerning form, structure, relation, and coherence date back to the work of Pythagoras (570–480 BCE). In the East, thought concerning relationship, transformation, and dynamic coherence goes back to the classic text I Ching (circa 780 BCE), whose title can be translated as "Classics of Transformations" or "Book of Changes."

Consistent with this background in holistic thought, ancient forms of medicine are typified by this form of thinking. For instance, in Traditional Chinese Medicine, all diseases are attributed to two possible causes, either detrimental emotion or discoherence between the individual and their environment. If we analyze this on a more fundamental level, because unhealthy emotion can also be considered a form of discoherence with society, discoherence is considered the fundamental cause of all problems.

The great success of rationalism and the industrial revolution have seen thinking characterized by materialism and reductionism come to dominate the world. Instead of coherence, this style of thinking focuses on conquest: conquering other nations, conquering nature, conquering bacteria, and conquering cancer. The preponderance of this mode of thought led to holistic thinking being largely ignored for centuries.

Nowadays, humanity and the rest of the world has started to bear increasing consequences of this conquering-focused thinking. Millions have perished as a result of the countless wars this mode of thinking creates. Its costs can also be seen in the ever-increasing damage being done to the natural environment. In addition to the damage done by industrial pollutants, awareness of the adverse effects of the pervasive use of disease-conquering chemical medical remedies is growing. Consequently, more people are reconsidering the importance of environmental protection, ecological balance, holistic thinking, and coherence in the world, in society, and inside our bodies.

From Standard of Living to Quality of Life

History is littered with tales of wars and starvation. After World War II, people suffered from a great shortage of food, clothing, housing, and transportation, and so improving the standard of living by addressing these needs represented the most important matter. Following sixty years of relative peace, the standard of living has improved dramatically in many countries. Instead of a shortage of food, one of the biggest problems facing many societies is obesity.

People have become more aware that standards of living should not be the only criterion by which our lives are evaluated. Increasing attention is now being paid to the goal of quality of life or wellness. Material comforts, energy abundance, and reductionism offer only limited future potential benefits to quality of life. As such, people have started to reconsider aspects of interrelation, structure, form, transformation, and coherence that were considered deeply in ancient wisdom.

Reductionist conquering thinking in medicine has saved countless lives, but it does not provide the comprehensive means to achieve wellness. While science has peered deeply to reveal all the constituents in the body, an understanding of the complexities of the relationships among the elements is lacking. We possess the full schematic of the human body as a machine but cannot fully comprehend the role played by emotion and consciousness. All the individual elements as well as consciousness have to achieve coherent cooperation in order for real wellness to be achieved. Developing means of achieving this coherence has become a new challenge for modern science.

Coherence Is Dynamic

As discussed, modern science has developed the means to address the formidable challenge of measuring and evaluating coherence in the specialized application of Hilbert space called the coherence pyramid, which allows quantitative evaluation of a patient's state relative to perfect coherence.

To effectively apply this technology in health care, three important points must be kept in mind. First, no real person is able to achieve the ideal state at the peak of the pyramid, although some highly coherent individuals might get close. Second, coherence is a dynamic concept rather than a static one. In other words, the state of a real person is always vacillating between concentration and relaxation (plate 17 in color plate section). It is normal, healthy, and even necessary to swing to the chaotic state, for example in deep relaxation, and the crystal state, when working or driving.

Health care systems should maintain awareness of the dynamic influence of lifestyle rhythms. For example, someone spending too much time in the crystal state, required for hard work, and consequently harboring accumulated stress in the body, should consider the necessity of relaxation before it is too late. Conversely, some elderly or wealthy people with too little focus in their lives who consequently stray too far toward the ideal chaotic state would benefit from establishing and pursuing some goals. This would help them come closer to achieving dynamic balance between concentration and relaxation.

If an adverse state arises from an individual's situation in society, such as unemployment, financial trouble, an excessive workload, or dysfunctional relationships with family or the community, interventions are required to help regain balance. Programs to release stress and depression can be formulated to achieve this before the cumulative effect of discoherence causes more drastic health damage.

The third point is that it is unnecessary to be excessively concerned when our states stray a significant distance from the coherence state for a relatively prolonged period. People usually have a strong capacity to regain their health automatically once the cause of the discoherence is addressed. Unlike a machine, a living organism is not only capable of self-repair but also of regaining balance and coherence spontaneously. In this context, the green region in plate 18 (in color plate section) can be regarded as a kind of attractor, automatically attracting the state of the patient back toward a more beneficial balance.

It must be noted that the presence of a self-repairing mechanism in our bodies does not entitle us to perpetually allow damaging influences like continuously excessive workloads, alcohol consumption, smoking, family drama, or partying. There are limits beyond which our systems cannot recover.

The Incorrect Attractor: Stubborn Chronic Disease

Aside from acute health problems, the most common phenomenon is that the attractor moves to a detrimental position. This occurs as a result of accumulated discoherence; prolonged stress is an example of a potential cause for this. Instead of attracting the system toward a healthier position, it has the opposite effect, as seen in plates 19 and 20 (in the color plate section). In medical terminology, a patient in this state is usually described as suffering from a chronic disease.

The great challenge facing modern medicine is chronic disease. While Western medicine is effective in tackling acute medical problems, its success in addressing chronic conditions is limited because there are no specific pathogens for the reductionist approach to target, or the virus or bacterium in question is resistant to chemical remedies.

Another weakness of Western medicine in dealing with chronic disease is that it is, for the most part, based on single-factor thinking. Even with extensive medical tests, there are many cases where physicians are unable to find the underlying reason for the disease, even though the location of the problem is clear. In these cases, the problem is then attributable to a functional disorder rather than pathogenic disease. As such, while chemical remedies can alleviate pain and other symptoms, there is no medication to address the underlying problem. In these situations, single-factor thinking is unsuccessful.

The weakness of Western medicine and its associated reductionist thought in treating chronic conditions has spurred patients and doctors to review ancient medical systems in search of new remedies and methods. In addition to reviving ancient therapies, doctors have invented a variety of new methods of treatment. Most of these fall into the category of physiotherapy and provide physical stimulation to help the body regain coherence.

In addition to the challenges faced in treating chronic conditions, it is also difficult to measure these diseases objectively. With functional disorders, it is particularly difficult to move beyond subjective descriptions in order to gather quantitative scientific measurements. The recognition of the electromagnetic body enables the degree of coherence to be objectively quantified. This allows the status of a patient to be represented in the coherence pyramid.

Chronic suffering is usually stubborn. Complementary therapies, including massage, music, acupuncture, homeopathy, and psychological counseling can often assist a patient to regain coherence temporarily, but after a brief respite, the patient returns to the previous unhealthy state. In the context of the coherence pyramid, this can be attributed to an incorrect attractor drawing the patient back to the unhealthy state. Essentially, chronic disease treatment represents a prolonged tug-of-war between the wrong attractor and the collaboration of the doctor and patient. While the protracted nature of the struggle is usually frustrating for all involved, it must be noted that chronic disease is usually the consequence of accumulated discoherence over a long period of time, and consequently cannot be rectified overnight.

Objective measurement and evaluation of coherence does not offer an easy cure to the problem of chronic suffering. It does, however, provide doctors and patients with a clear picture of where they are (plate 21 in the color plate section). This allows doctors to devise the optimal strategy to help the patient, and enables the patient to appreciate the degree of patience and persistence required for success.

From Ancient Wisdom to Modern Science

Many practitioners of holistic forms of medicine are familiar with the tai-chi symbol (at right in fig. 15.6), a symbolic expression of dynamic balance. White symbolizes yang and black is yin. Yin and yang are philosophical concepts that can have many meanings in representing antagonistic entities. Yang, for example, can mean male, strong, powerful, bright, hard, or a governing political party. Yin would have the respective corresponding meanings of female, weak, feeble, dark, soft, and an opposition party.

In addition to opposing each other, yin and yang are also in a permanent dynamic balance. Any system with two opposing entities in dynamic balance can be expressed in terms of the waxing and waning of yin and yang on a clockwise-rotating tai-chi wheel. Consequently, the tai-chi symbol can also be used to symbolize the dynamic balance required between concentration and relaxation. This can be quantitatively represented in the coherence pyramid (fig. 15.6).

In the tai-chi symbol, as time passes, the white, representing yang, grows (phase 1 in fig. 15.6) until it reaches its maximum (phase 2). At the point where

Coherence in Medicine and Health Care

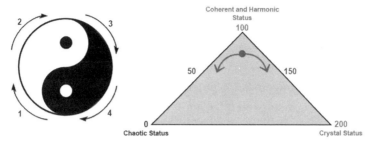

Figure 15.6. From the philosophical tai-chi to the practical coherence pyramid.

the yang is greatest, a small circle of black, representing the seed of the yin inside the yang, occurs. After this, the yang diminishes and the yin grows (phase 3), until its maximum, where the white seed of yang appears inside the yin (phase 4). Then the cycle begins again.

From the viewpoint of physics, this behavior is consistent with vibration. Consider the movement of a pendulum, which can be mathematically shown in a sine curve. This sine curve also serves as the basis for all waves, including the acoustic waves in music and the electromagnetic waves inherent in the acupuncture system of Chinese medicine, the chakra system of Indian medicine, and homeopathy, as discussed in this book.

In addition to being admired by alternative healing practitioners in the West, the tai-chi symbol was also appreciated by some modern physicists, including Niels Bohr, the founder of quantum physics, who won the 1922 Nobel Prize in physics for investigating the structure of atoms and radiation emanating from them; and his protégé, Werner Heisenberg, most famous for the uncertainty principle, who won the 1932 Nobel Prize for the creation of quantum mechanics. Bohr's appreciation of the tai-chi symbol was seen when he was awarded the Order of the Elephant by the Danish government for winning the Nobel Prize; he designed his own coat of arms with the tai-chi at its center.

Figure 15.7. Niels Bohr (left) and Werner Heisenberg (right), the founders of quantum physics.

Bohr and Heisenberg belonged to a pioneering generation of physicists who ventured from the visible mechanical world to the invisible atomic level. They found that behavior at the quantum level is beyond the descriptive ability of the language and concepts in our everyday world. Many, including the great physicist Einstein, could not accept the strange behavior of quantum physics. Physicists working in this area faced not only the problem of acceptance but also challenges posed by the limitations of our ability to conceptualize—challenges akin to those facing the scientists in the world of the blind in chapter 1.

Bohr and Heisenberg worked at the frontier of science, and their research made them painfully aware of our limitations. Our senses are unable to perceive the atom and electron directly; our language, developed in our macroworld, is insufficient to describe the strange behavior of the invisible and untouchable quantum world; and our minds, fashioned by the reality of our world, are ill equipped to understand its counterintuitive reality.

The challenges we face in understanding the electromagnetic body, the invisible rainbow and inaudible music in living systems, stem from the same limitations. People have come to accept the existence of the ethereal world of the atom and the electron and even take it for granted. Modern scientists also acknowledge the counterintuitive behavior of the quantum realm and apply it in technology without considering its contentious beginnings.

It is difficult to write a book describing a new world that is beyond our senses, beyond the ability of our language, and even sometimes beyond our imagination. It is also difficult to read such a book, but I hope it has been an enjoyable read.

I like to think that Pythagoras, Lao Tzu, and other ancient sages would be pleased if they knew that their wisdom still inspired and enlightened the thinking of scientists today. Similarly, I believe they would be happy to know that their philosophy concerning form, relation, dynamic balance, and coherence is now within the reach of modern science, offering rational and quantitative understanding, and applications for the health care of humanity.

EPILOGUE

CONSCIOUSNESS, SPIRIT, AND CONSCIENCE IN SCIENCE

Einstein once described the theory of relativity by saying, "When you sit with a nice girl for two hours you think it's only a minute, but when you sit on a hot stove for a minute you think it's two hours. That's relativity."[1]

Objectivity and Subjectivity

This unusual explanation is usually considered humorous, but on a deeper level it embodies a major challenge faced by Einstein and modern frontier scientists: the fundamental challenge of objectivity. Since its inception, modern science has espoused the axiom that is absolutely objective and independent of the presence of people. This belief is reflected in Newton's framework of time and space. Both are considered to be absolute and unaffected by whether they are being observed or not.

Einstein first introduced the fiend of the subjective observer into the temple of science. From then on, time and space were no longer absolute but relative and dependent on the state of the observer. In the special theory of relativity, time and space depend on the speed of movement of the observer. As the observer's speed increases, time slows down and distances are reduced. The general theory of relativity introduces the acceleration of the observer. We are the observers that warp time and space and confound the revered objectivity of science.

Both Einstein's theories of relativity referenced simplified movement in a straight line. Given that the simple movement of an observer can have such profound implications for time and space, the influence of the observer's state of mind also needs to be considered. Developing the theory of relativity to encompass the study of complex systems, such as living systems and the body-mind system, poses a complex challenge, and a solution was not possible during Einstein's lifetime. The challenge of developing the theory of relativity to address these requirements is left to those who follow in his footsteps.

The problem of the influence of the observer was also encountered by quantum physicists, including Bohr and Heisenberg. They found that the atom, the instrument, and the observer are an inseparable system. Heisenberg expressed this as: "What we observe is not nature itself, but nature exposed to our mode of questioning."[2] It is impossible to establish a scientific theory that is completely objective—it is always relative to the subject's interaction with the people observing.

Interestingly enough, the problem of the inseparability of observers and the objective world was discussed by the famous Zen Buddhist Huineng about 1,500 years ago. As an illiterate underprivileged youth, Huineng joined a temple in a low position, but his abilities were recognized by the fifth Great Master of Zen Buddhism, who secretly gave his *kasaya* (Buddhist robe) to Huineng and chose him as a successor. The master knew that Huineng would be in great danger from jealous senior apprentices and asked him to flee the temple for his own safety.

After living in seclusion for fifteen years, Huineng participated in a large conference on Buddhism, where he heard two Buddhists arguing about a waving flag. One said, "Look, the flag is moving," and the other responded, "No, you are wrong—the wind is moving." Huineng said, "It is not the flag or the wind; it is the mind that is moving." Hearing his comment, the other participants realized he was the sixth Great Master of Zen Buddhism, who had been missing for fifteen years.

Huineng's response became encapsulated in the fundamental principle of Zen Buddhism that "nothing is reality, but mind." It is worth noting that this principle is the polar opposite of materialism. Let us compare it to Lenin's definition of *material*, discussed in chapter 2, which he formulated to rescue materialism from the challenge of consciousness: "Material is objective existence which is independent from the consciousness."[3] This new definition included energy in the realm of matter but did not provide a definition for consciousness or mind. In contrast, Zen Buddhism holds that mind, or consciousness, is much more fundamental than matter, energy, or field.

For a long time, the terms *mind* and *consciousness* were not considered part of science. Anyone talking about mind or consciousness in a scientific context would be regarded as superstitious and dismissed, but the last twenty years have seen dramatic changes in the way that mind and consciousness are considered. The terms have lost their stigma and their use has spread in many fields of science, but despite wide discussion, no one can say exactly what they mean.

The major change in scientific acceptance of consciousness was initiated by brain research projects in the United States. In the late 1980s, the U.S. government launched a relatively ambitious national program in brain research called "Decade of the Brain." The thinking was that, given the rapid development of biology, particularly molecular biology, the only remaining unclear corner of the body was the brain. The government believed that if enough money was dedicated to the research, all the biological questions about the brain could be answered during the twentieth century.

The initial goal of the project was simple. The brain was essentially considered a computer, and given a sufficiently powerful computer, which existed at the time, and the appropriate programming, it would be possible to imitate any ability of intelligence. After a couple of years, an insurmountable barrier appeared: no computer possessed self-awareness. In other words, while every child possessed the awareness of "I am," computers did not have the consciousness to do so.

Scientific stigma notwithstanding, renowned physicist Erwin Schrödinger (fig. 16.1) discussed the problem of consciousness from the viewpoint of science over half a century ago. Physics students are familiar with his name because the Schrödinger equation forms the basis of quantum physics. Similarly, even nonphysicists are familiar with the analogy of Schrödinger's cat. But relatively few scientists are aware that in 1944 he wrote a book titled *What Is Life?*, with the final chapter titled "What Is I?"

Figure 16.1. Erwin Schrödinger, the first scientist to seriously consider what life is from the viewpoint of physics.

In the book, Schrödinger predicted an "aperiodic crystal" that carries all the genetic information for living systems. He used data on mutation rates under alpha radiation and also calculated the size of an individual gene. It is particularly impressive that the size he calculated is exactly the size of the triplet code in DNA. It is evident that the aperiodic crystal he speculated about is the subsequently discovered DNA, which plays a profound role in biology today.

In the last chapter of his book, Schrödinger discussed the definition of *I*. He concluded that *I* is merely a singular memory. As such, the recognition of the world is only the accumulation of memory. The common sense of the world represents the consensus of many individual memories. In other words, science is only the agreement of many individual observations, namely the subjective experiences of many observers. Consequently, science represents only a common agreement of many individual experiences, not the objective truth of the world. That being said, it is still the most elegant body of knowledge created by humanity.

The limitations of objectivity in science have been discussed by Heisenberg and many other outstanding physicists, including Bohr and Einstein. Bohr and Heisenberg found that the observer, the instrument, and the microworld are actually an inseparable system. Heisenberg formulated this mathematically in the uncertainty principle, which earned him a Nobel Prize. Einstein's humorous statement about relativity alludes to the idea that the interaction between our consciousness and the rest of the world is dependent not only on the observer's movement but also on his or her physiological state.

The absence of pure objectivity is not only a fundamental problem for science; it also represents a perpetual challenge for art and aesthetics. For instance, beauty is neither purely objective nor purely subjective—it is an interaction between object and consciousness.

Consciousness, Spirit, and Conscience

This book's discussion has focused on the dissipative structure of electromagnetic fields and its ability to bring the ancient wisdom of acupuncture, chakras, and many other holistic forms of medicine into modern science. In doing so, it also considered the objective measurement and quantitative evaluation of harmony, which also enables the wisdom in ancient philosophy and art to be modernized. However, the discussion did not extend to serious consideration of life, consciousness, spirit, and conscience from the perspective of modern science. Frankly, the nature of life and spirit currently exists outside rigorous scientific understanding. That being said, the names of some branches of science, such as biology and psychology, proclaim that they have knowledge of life and soul.

Ancient Indian philosophy holds that an individual has seven bodies, or seven body-levels. Let us consider science's present and potential future understanding of the human body, mind, and consciousness within the context of this framework. The solid structure of the body, including the organs, tissues, cells, and the molecules, belongs to the first level, namely the chemical body.

At the level of the chemical body, the scientific law of the conservation of mass applies: mass is neither created nor destroyed in any normal chemical reaction. It means that the atoms in a body will continue to exist eternally, assuming they don't become involved in a nuclear interaction, even after the person had died. They will be recycled into the world, some forming other people's bodies. In some ways this can be considered a kind of reincarnation, as discussed thousands of years ago in Buddhism and in the Christian Bible.[4]

The dynamic dissipative structure of electromagnetic fields in living systems belongs to the second level of body, the electromagnetic body, usually inseparable from the chemical body; they are interdependent and interact continuously while an individual is alive. At this level, the law of conservation of energy applies. While energy changes from one form to another, the amount of energy remains constant. This essentially amounts to a quantitative law for the reincarnation of energy.

Albert Einstein discovered the law of transformation between matter and energy. In doing so, he merged material reincarnation and energy reincarnation into a single reincarnation principle. The reincarnation of energy poses a small complication: unlike matter, energy always transforms from higher forms to lower forms. For example, electrical energy will be dissipated as heat energy; the reverse process never occurs automatically. It seems that energy is not comprehensively conserved, at least in terms of its form or structure.

To describe the irreversible process of energy transformation quantitatively, we look to the second law of thermodynamics, which holds that overall, entropy will always irreversibly increase. Entropy means the degree of disorder, in essence the opposite of structure. Structure will irreversibly decrease and eventually disappear. In other words, there is a perpetual trend toward degeneracy in structure.

In the 1970s, almost one hundred years after the discovery of the second law of thermodynamics, the opposite phenomenon, the dissipative structure, was discovered. It showed that as the disorder of energy increases, a new order, in the form of dissipative structure, simultaneously arises. They counteract each other and suggest the possibility of a new transformative relationship between order in energy and order in dissipative structure. In other words, the permanent degeneration of structure might be balanced by the newly developed structure.

Similar to the way that Einstein's law merged the transformation of matter and energy, the possible transformative relationship in structure could offer the opportunity for a higher level of conservation law that merges the conservation of matter, energy, and structure. It would offer a system of general reincarnation for all three aspects. The subsequent higher levels alluded to in the traditional Indian understanding of the body might also possess additional conservation laws describing life, consciousness, spirit, and conscience. Of course, significant additional progress is required before these can be discovered.

Perhaps conscience exists in the highest level of a human body. Throughout the history of science, the problem of scientific development applied for good or for evil has troubled scientists. This raises an even more fundamental question: what is good and what is evil? For instance, the decades-old debate as to whether nuclear energy is good or evil still continues. Similarly, modern developments like genetic engineering pose the same question.

It is worth noting here that conscience is somewhat opposed to Darwinism in that it is at odds with the fundamental Darwinian principle of ruthlessly fighting for survival. A Russian friend of mine said; "If there is some truth in Darwinism, it is at most a half theory, like the second law of thermodynamics. There must be another half. If we say Darwinism is the ugly and cruel half of evolution, there must be a beautiful and virtuous half of evolution." If this is true, this other half must provide some scientific and quantitative criteria to differentiate what is good and what is evil.

The Limitation of Our Senses and Language

Perhaps the goal of incorporating consciousness into scientific theory is too ambitious. The limitations of our senses, discussed in chapter 1, provide a significant impediment. While we can construct instruments to extend the abilities of our senses, such as telescopes and microscopes, these act as transformers, thereby introducing distortions. The more instruments we use, the more distortions arise. In the end, we would approach a limitation of distortion, which would be the boundary of our recognition.

We have known quite a few boundaries in the textbooks of physics. The speed of light constant is a typical boundary for us, as is Planck's constant. There are many others. In fact, most constants are the boundaries of our limitations.

Besides the limitations of our sense organs, we have serious problems in the limitations of our language. Our daily language was developed with only the experience of the first level of body, the physical body. We have seen in this book how difficult it is to describe the second level of body, the electromagnetic body, with the language of the physical body. Proceeding into higher levels of body would make using our daily language to describe them even more problematic. The great philosopher Lao Tzu clearly expressed the limitation of language when he said, "The word that can be spoken is not the real Word. The name that can be named is not the real Name."[5] It is worth noting that the "Word" Lao Tzu is referring to can be regarded as the same as in the Christian Bible's "In the beginning was the Word,"[6] as well as "And the Word was made flesh, and dwelt among us, full of glory and truth."[7]

Therefore, the more we know the limitations of our language, our senses, and our intelligence, the closer we are to glory and truth.

AFTERWORD

ENDLESS EXPLORATION

The first part of this book, "The World of the Blind," originated from a long letter that I wrote to my friend Xu-liang Hu more than forty years ago. At the time, we were both very young, full of dreams and ambitions, and quite critical of authority.

It was a crazy time in China, officially referred to as the Great Proletarian Cultural Revolution, but it was actually a civil war waged for ten years until the death of Mao Ze-dong. Incredibly, almost all books, with the exception of those written by Marx, Engels, Lenin, Stalin, and Mao, were forbidden and burned. Every university was closed, and all young educated people were sent to the countryside to receive "reeducation by poor and lower-middle-class peasants." In essence it was punitive brainwashing of educated people.

There were still some young people who longed for knowledge and secretly read books, which were very difficult to find at the time. During vacations, they returned to their homes in the cities and gathered to exchange books and their opinions and insights from their readings. Sometimes, a reader would deliver a speech—this would prompt enthusiastic and deep discussion within the group.

These readers had quite different interests and studied different disciplines. As such, philosophy represented a common topic for all of us, and we read widely, from Aristotle to Immanuel Kant, from Gautama Buddha to Lao Tzu. We greatly enjoyed the opportunity for our minds to venture into the worlds of the philosophers, to learn and to absorb their teachings like dry sponges.

The philosophies we encountered were diverse, and their outlooks differed widely and were often contradictory to one another. Surprisingly, however, all of them successfully presented a beautiful, self-consistent, and seemingly perfect explanation for the world that we inhabit. Obviously, it is impossible that all of them were telling the absolute truth.

This led us to wonder whether absolute truth exists. Gradually, we arrived at the notion that none of them tell us the real truth, that perhaps each of them only conveys a partial truth, possibly even distorted, but describes their theory as the whole truth. The situation is somewhat akin to frogs sitting at the bottom of different wells; each of these frogs sees a sky that is circular and perfect. Each then describes their round and perfect sky and respectively establishes a special and seemingly perfect philosophical system. In this context, all the great philosophers in history were brilliant frogs, sitting at the bottoms of various wells, describing the perfect sky that they carefully observed.

This consideration of philosophers provoked within me a serious conundrum. If our insights reflected the situation of philosophers, what was the situation of us scientists? One night I had some inspiration and poured it into a long letter to Hu about the hypothetical situation of the scientists in the world of the blind.

I am truly grateful to this discussion group because I have continuously benefited from it during my career in a frontier area of science. This letter about the scientists in the world of the blind has guided my work, and eventually helped me to discover the breathtakingly beautiful invisible rainbow, the inaudible music, and much more. In some way, this book is a record of four decades of spiritual journey. It is my hope that I have been able to convey enough of a sense of the beautiful landscapes along my journey for you to be able to appreciate them with me.

In reality, science is a creation made by human beings like you and me, who are far from holiness and absolute truth. Science is the same: as demonstrated in the story of the scientists in the world of the blind, it is neither perfect and holy, nor the perfect representation of truth. Science is merely an adventurous enterprise and an attempt to understand the infinite world with our finite limitations: our limited senses and our limited rational thinking, reasoning, and languages.

That being said, modern science is the most rigorous, beautiful, even splendid knowledge system ever constructed by humanity. In its purest form, scientific research is a profound exploration into unknown worlds for the benefit of all humanity. It requires courage, persistence, dedication, and humility to tirelessly search for truth.

The robust development of science requires a large number of scientists to follow the example of Max Planck, founder of quantum physics. Einstein, in the introduction he wrote for Planck's book *Where Is Science Going?*, described him as one of the few worshipers who would remain in the Temple of Science should an angel of God descend and drive out the lesser scientists, who under different circumstances might become politicians or captains of industry.[1]

This selfless devotion to the spirit of science is required for its healthy development. Therefore, let us always seek to follow this spirit and humble ourselves in order to find more pretty shells and smooth rocks to examine on the shores of the ocean of truth.

CHANGLIN ZHANG

DECEMBER 2015, IN GIESSEN, GERMANY,

THE SMALL UNIVERSITY TOWN WHERE WILHELM RÖNTGEN

AND JUSTUS VON LIEBIG ONCE WORKED

NOTES

Foreword

1. Joachim-Ernst Berendt, *Nada Brahma. Die Welt ist Klang*, Reinbek, 1983, S. 28 ff; published in English as Joachim-Ernst Berendt, *Nada Brahma: The World Is Sound: Music and the Landscape of Consciousness* (Rochester, VT: Destiny Books, 1987).

2. Alfred A. Tomatis, *Der Klang des Lebens. Vorgeburtliche Kommunikation—die Anfänge der seelischen Entwicklung*, Reinbek, 1987, S. 173 ff.

3. Berendt, *Nada Brahma*, a.a.O. S. 226.

Preface

1. David Brewster, *Memoirs of the Life, Writings, and Discoveries of Sir Isaac Newton*, 1855, vol. 2, ch. 27.

1. Revisiting "Blind Men Study an Elephant"

1. Electromagnetic waves are composed of electric and magnetic fields. Light waves, radio waves, X-rays, ultraviolet rays, infrared rays, and microwaves are all examples of electromagnetic waves.

2. A nanometer (nm) is one-billionth of a meter. To visualize this, if you took 1 millimeter, which equals about four-hundredths of an inch, and divided it into 1,000 parts, and took one of those parts and divided it by 1,000 again, you would have a nanometer.

3. We perceive different wavelengths of light as different colors. For example, we perceive light of 700–635 nm wavelengths as red, 590–560 nm wavelengths as yellow, and 450–400 nm wavelengths as violet.

2. Spiritualized Physics and Materialized Psychology and Biology

1. Albert Szent-Györgyi, *Nature of Life: A Study on Muscle* (New York: Academic Press, 1948), 9.

2. Examples of such considerations can be found in publications at the Spirituality and Psychiatry Special Interest Group of the Royal College of Psychiatrists in the United Kingdom (www.rcpsych.ac.uk/college/specialinterestgroups/spirituality.aspx).

3. The four fundamental forces in nature are gravity, the electromagnetic force, the strong nuclear force, and the weak nuclear force.

4. Werner Heisenberg, *Physics and Philosophy: The Revolution in Modern Science* (New York: Harper, 1958), 30.

5. Hebrews 11:3.

6. 2 Corinthians 4:18.

7. Vladimir Lenin, *Materialism and Empirio-criticism: Critical Comments on a Reactionary Philosophy* (Moscow: Zveno, 1909), ch. 5.8.

3. Inaudible Music and the Invisible Rainbow around Us

1. Werner Heisenberg, *The Physical Principles of the Quantum Theory* (New York: Dover, 1950), 11.

4. Major Changes in the Medical Market

1. David M. Eisenberg, Ronald C. Kessler, Cindy Foster, Frances E. Norlock, David R. Calkins, and Thomas L. Delbanco, "Unconventional Medicine in the United States—Prevalence, Costs, and Patterns of Use," *New England Journal of Medicine* 328 (January 28, 1993): 246–52, doi:10.1056/NEJM199301283280406.

2. Applied kinesiology is an alternative medicine technique that claims to be able to diagnose and recommend treatment by testing muscles responses for strength and weakness. Bio-oxidative therapies are various practices in which oxygen, ozone, or hydrogen peroxide are administered, either via gas or water, to promote health. Qigong is an umbrella term that covers several different Chinese systems of energy cultivation and healing, including breathing techniques, movement, and meditation. Iridology is an alternative diagnosis technique that examines the iris to discern details about a patient's health.

3. Tillman Durdin, "Tientsin Doctor of Acupuncture Says His Needles Cure Many Ills," *New York Times*, April 28, 1971; James Reston, "Now, About My Operation in Peking; Now, Let Me Tell You about My Appendectomy in Peking...," *New York Times*, July 25, 1971.

4. Boyce Rensberger, "U.S. Doctors Are Skeptical of Acupuncture in Treatment of Purely Physical Diseases," *New York Times,* October 7, 1971.

5. The concept of synergetics is well explained in Haken's book, *The Science of Structure: Synergetics* (New York: Van Nostrand Reinhold, 1984). Dissipative structures are discussed in detail in chapter 7. Although the uncertainty principle applies to certain different pairs of properties for particles on the quantum scale, the most common example is for the momentum and position of a particle. Essentially, the more precisely you know a particle's position, the more uncertainty you will have about its momentum, and vice versa.

5. Queen Victoria Studies TV

1. This view was expressed by Arthur Taub, Director of the Neurosurgical Research Laboratory and the Pain Diagnostic and Therapeutic Division of the Neurosurgery Section at Yale University, in his letter "Acupuncture: U.S.-Funded Research Is Premature," published in the *New York Times,* August 8, 1972.

2. Boyce Rensberger, "Acupuncture Likened to Placebo," *New York Times,* June 19, 1975.

3. Xiang-Long Hu, et al., *Modern Scientific Research in Acupuncture Channels and Collaterals in Traditional Chinese Medicine* (Beijing: People's Hygiene Publishing House, 1990) (in Chinese; ISBN 7-117-01415-6).

6. Blind Scientists Discover the Rainbow

1. Hu, *Modern Scientific Research,* 189.

2. C.-E. Overhof, "Über das elektrische Verhalten spezieller Haubezirke" (dissertation, Fakultät für Maschinenwesen, Karlsruhe Institute of Technology, 1960).

3. Author's personal communication with Konstantin Zioutas, and Changlin Zhang, "Skin Resistance versus Body Conductivity: On the Background of Electronic Measurement on Skin," *Frontier Perspectives* 11:2 (2002), 15–25.

4. Z. X. Zhu, et al., *Biophysics for Acupuncture* (Beijing: Beijing Publishing House, 1989), 323 (in Chinese).

5. Hiroshi Motoyama, "Before Polarization Current and the Acupuncture Meridians," *Journal of Holistic Medicine* 8 (1 and 2: 1986), 15–26, https://www.cihs.edu/wordpress/wp-content/uploads/2012/01/Before-Polarization-current-and-acupuncture-meridians.pdf.

7. A New Continent in Science

1. U.S. Patent and Trademark Office, Patent Application Database, http://goo.gl/QzOSeB.

8. Standing Waves and Wave Superposition

1. Author's personal communication with Hans-Jürgen Stöckmann.

9. Wireless Communication inside a Body

No notes.

10. Powerful Resonance

1. Eric Allen, *Airliners in Australian Service* (Western Creek, Australia: Aerospace Publications, 1995), 155.

2. Claude Elwood Shannon and Warren Weaver, *The Mathematical Theory of Communication*. Urbana, IL: University of Illinois Press, 1949.

3. Wiener, Norbert. *Cybernetics; or, Control and Communication in the Animal and the Machine*. New York: J. Wiley, 1948.

4. Cyril W. Smith, "Homeopathy: How It Works and How It Is Done," Hpathy Ezine, January 16, 2008, http://hpathy.com/scientific-research/homeopathy-%E2%80%93-how-it-works-and-how-it-is-done-1/.

5. See Brenda J. Dunne, Roger D. Nelson, and Robert G. Jahn, "Operator-Related Anomalies in a Random Mechanical Cascade," *Journal of Scientific Exploration* 2:2 (1988), 155–79, doi:0892-3310/88; and Princeton Engineering Anomalies Research, http://www.princeton.edu/~pear/publications.html.

6. Paul Pojman, "Ernst Mach," *The Stanford Encyclopedia of Philosophy* (winter 2011 ed.), Edward N. Zalta, ed., http://plato.stanford.edu/archives/win2011/entries/ernst-mach/.

7. Heisenberg, *Physics and Philosophy*, 30.

8. Hebrews 11:1.

11. The Mysterious Aura

1. A spectrophotometer measures the intensity of the different wavelengths of electromagnetic waves entering it, telling researchers the amount of electromagnetic waves at each wavelength. In this context, "absorption spectra" refer to a graphical representation of the intensities of the different wavelengths of electromagnetic waves in the sample. The intensity of each wavelength is shown on a vertical scale.

2. See Chris Cook, Andrew S. Powell, and Andrew C. P. Sims, eds., *Spirituality and Psychiatry* (London: RCPsych, 2009); and Royal College of Psychiatrists, Spirituality and Psychiatry Special Interest Group, "Publications Archive," http://goo.gl/SrMWyk.

3. Stefan Lovgren, "Dogs Smell Cancer in Patients' Breath, Study Shows," National Geographic News, January 12, 2006, http://news.nationalgeographic.com/news/2006/01/0112_060112_dog_cancer.html.

12. Facing Complex Systems

1. Karl R. Popper, *Conjectures and Refutations: The Growth of Scientific Knowledge* (New York: Basic Books, 1962), ch. 1.

2. Lao Tzu, Tao Te Ching, ch. 1.

13. How Much Beauty Is There in a Ballet?

1. Changlin Zhang and Fritz-Albert Popp, "Log-Normal Distribution of Physiological Parameters and the Coherence of Biological Systems," *Medical Hypotheses* 43:1 (1994), 11–16; Changlin Zhang, "Mathematical, Physical and Physiological Background of Normal Distribution, Delta Distribution and Log-Normal Distribution," *International Journal of Modeling, Identification and Control* 5:3 (2008), 200–4.

14. Measuring the Invisible Rainbow

No notes.

15. Coherence in Medicine and Health Care

No notes.

Epilogue

1. "Einstein Is Found Hiding on Birthday: Busy with Gift Microscope," *New York Times*, March 15, 1929.

2. Heisenberg, *Physics and Philosophy*, 30.

3. Lenin, *Materialism*, ch. 5.8.

4. John 9:1–4.

5. Lao Tzu, Tao Te Ching, ch. 1.

6. John 1:1.

7. John 1:14.

Afterword

1. Albert Einstein, Prologue to Max Planck, *Where Is Science Going?* Trans. James Vincent Murphy (New York: W. W. Norton, 1932), 7.

BIBLIOGRAPHY

Allen, Eric. *Airliners in Australian Service*. Western Creek, Australia: Aerospace Publications, 1995.

Bentov, Itzhak. *Stalking the Wild Pendulum: On the Mechanics of Consciousness*. New York: E. P. Dutton, 1977.

Berendt, Joachim-Ernst. *Nada Brahma: The World Is Sound: Music and the Landscape of Consciousness*. Rochester, VT: Destiny Books, 1987.

Bohr, Niels. *Atomic Theory and the Description of Nature*. New York: Macmillan, 1934.

Brewster, David. *Memoirs of the Life, Writings, and Discoveries of Sir Isaac Newton*, 1855.

Capra, Fritjof. *The Tao of Physics: An Exploration of the Parallels between Modern Physics and Eastern Mysticism*. Berkeley, CA: Shambhala, 1975.

Cook, Chris, Andrew S. Powell, and Andrew C. P. Sims, editors. *Spirituality and Psychiatry*. London: RCPsych, 2009.

Dunne, Brenda J., Roger D. Nelson, and Robert G. Jahn. "Operator-Related Anomalies in a Random Mechanical Cascade." *Journal of Scientific Exploration* 2:2 (1988), 155–79, doi:0892-3310/88.

Eddington, Arthur Stanley. *The Nature of the Physical World*. New York: Macmillan, 1928.

Eisenberg, David M., Ronald C. Kessler, Cindy Foster, Frances E. Norlock, David R. Calkins, and Thomas L. Delbanco. "Unconventional Medicine in the United States—Prevalence, Costs, and Patterns of Use." *New England Journal of Medicine* 328 (January 28, 1993): 246–52, doi:10.1056/NEJM199301283280406.

Gibran, Kahlil. *The Forerunner, His Parables and Poems*. New York: A. A. Knopf, 1920.

Haken, Hermann. *The Science of Structure: Synergetics*. New York: Van Nostrand Reinhold, 1984.

Heisenberg, Werner. *The Physical Principles of the Quantum Theory*. New York: Dover, 1950.

———. *Physics and Philosophy: The Revolution in Modern Science.* New York: Harper, 1958.

Hu, Xiang-Long, et al. *Modern Scientific Research in Acupuncture Channels and Collaterals in Traditional Chinese Medicine.* Beijing: People's Hygiene Publishing House, 1990 (in Chinese; ISBN 7-117-01415-6).

Kaptchuk, Ted J. *The Web That Has No Weaver: Understanding Chinese Medicine.* New York: Congdon & Weed, 1983.

Lenin, Vladimir. *Materialism and Empirio-criticism: Critical Comments on a Reactionary Philosophy.* Moscow: Zveno, 1909.

Longo, Giuseppe. *Information Theory: New Trends and Open Problems.* New York: Springer, 1975.

Lovgren, Stefan. "Dogs Smell Cancer in Patients' Breath, Study Shows." National Geographic News. January 12, 2006. http://news.nationalgeographic.com/news/2006/01/0112_060112_dog_cancer.html.

Maciocia, Giovanni. Foreword to *The Vital Meridian: A Modern Exploration of Acupuncture,* by Alan Bensoussan. Melbourne: Churchill Livingstone, 1991.

Motoyama, Hiroshi. "Before Polarization Current and the Acupuncture Meridians." *Journal of Holistic Medicine* 8 (1 and 2: 1986), 15–26. https://www.cihs.edu/wordpress/wp-content/uploads/2012/01/Before-Polarization-current-and-acupuncture-meridians.pdf.

Overhof, C.-E. "Über das elektrische Verhalten spezieller Haubezirke." Dissertation, Fakultät für Maschinenwesen, Karlsruhe Institute of Technology, 1960.

Planck, Max. *Where Is Science Going?* Translated by James Vincent Murphy. New York: W. W. Norton, 1932.

Pojman, Paul. "Ernst Mach." *The Stanford Encyclopedia of Philosophy.* Winter 2011 edition. Edited by Edward N. Zalta. http://plato.stanford.edu/archives/win2011/entries/ernst-mach/.

Popp, Fritz-Albert, and Ulrich Warnke, editors. *Electromagnetic Bio-Information.* Baltimore: Urban & Schwarzenberg, 1989.

Popper, Karl R. *Conjectures and Refutations: The Growth of Scientific Knowledge.* New York: Basic Books, 1962.

Schrödinger, Erwin. *What Is Life? The Physical Aspect of the Living Cell.* Cambridge, UK: Cambridge University Press, 1944.

Shannon, Claude Elwood, and Warren Weaver. *The Mathematical Theory of Communication.* Urbana, IL: University of Illinois Press, 1949.

Smith, Cyril W. "Homeopathy: How It Works and How It Is Done." Hpathy Ezine. January 16, 2008. http://hpathy.com/scientific-research/homeopathy-%E2%80%93-how-it-works-and-how-it-is-done-1/.

Stapp, Henry P. *Mind, Matter, and Quantum Mechanics.* New York: Springer-Verlag, 1993.

Szent-Györgyi, Albert. *Introduction to a Submolecular Biology.* New York: Academic Press, 1960.

———. *Nature of Life: A Study on Muscle.* New York: Academic Press, 1948.

Wiener, Norbert. *Cybernetics; or, Control and Communication in the Animal and the Machine.* New York: J. Wiley, 1948.

Zhang, Changlin. "Mathematical, Physical and Physiological Background of Normal Distribution, Delta Distribution and Log-Normal Distribution." *International Journal of Modeling, Identification and Control* 5:3 (2008), 200–4.

———. "Skin Resistance versus Body Conductivity: On the Background of Electronic Measurement on Skin." *Frontier Perspectives* 11:2 (2002), 15–25.

Zhang, Changlin, and Fritz-Albert Popp. "Log-Normal Distribution of Physiological Parameters and the Coherence of Biological Systems." *Medical Hypotheses* 43:1 (1994), 11–16.

Zhu, Z. X., et al. *Biophysics for Acupuncture.* Beijing: Beijing Publishing House, 1989 (in Chinese).

INDEX

A

Achromatopsia, 138
Acu-points
  in the ear, 62
  movement of, 73–74, *Plates 8 and 9*
  shape of, 72
  size of, 71–72
Acupuncture. *See also* Acu-points; Meridians; Sensation propagation; Skin resistance measurements
  effectiveness of, 48
  history of, 46
  interference patterns and, 102–3, 114, *Plates 11 and 12*
  in plants, 66, 79, *Plate 6*
  popularity of, 48, 49
  puzzling phenomena of, 83
  research on, 47, 48, 55–62, 70–81
  scientific attitude toward, 47, 49
  third balance system hypothesis for, 67–68, 83
  wave guide channel hypothesis for, 68–69
Alternative medicine. See Complementary medicines
Ampère, André-Marie, 21
Andersen, Hans Christian, 136
Angers Bridge, 120
Annihilation reaction, 25
Anomalopia, 138
Antibiotics, 44
Antimatter, 25
Aristotle, 166, 168
Atom
  ancient theories of, 20
  chemical theory of, 21
  discovery of, 12, 20
  emptiness of, 25–26
  etymology of, 20
  models of, 23–24, *Plate 3*
Auras
  application of, in medicine, 140–41
  chemical, 140
  data analysis of, 144
  detecting, 143–44
  history of, 160
  internal, 142–43
  photographing, 140–41, *Plates 4 and 14*
  as religious concept, 133
  scientific evidence for, 137–38
  viewing, 31
Avogadro, Amedeo, 21
Axiomatic system, 153–54

B

Balance systems, 67–68
Ball-and-stick models, 21, 22, *Plate 2*
Beat frequency, 99, 185
Becker, R. O., 72
Belief
    power of, 132
    reality vs., 131–32
    as resonance, 132
    scientific acceptance as, 13
Bell, Alexander Graham, 105
Bentov, Itzhak, 104
Bernoulli, Daniel, 20, 24
Bhagavad Gita, 160
Bible, 164–65, 167, 207
Big Bang theory, 125, 154
Biofield, 16, 108–9
Biology
    etymology of, 15, 16
    materialism in, 15–16
    molecular, 22–23, 28, 108, 147
    open systems and, 92
    present situation of, 28–29, 113
    reductionism and, 147–48, 171
Body. *See also* Electromagnetic body
    chemical messenger system within, 104–5
    electrical messenger system within, 105
    energy distribution inside, 82, 100–103, *Plate 10*
    wireless communication within, 105–9
Bohr, Niels, 24, 119, 131, 199–200, 202, 204
Born, Max, 24
Boundary conditions, changing, 101–2, *Plates 11 and 12*
Brahe, Tycho, 19
Brahmanas, 160
Brain research, 203
Buddha, 133, 164
Buddhism, 26, 125, 161, 202, 205

C

Calvin, John, 167
Capra, Fritjof, 26, 52, 125
Carrier waves, 126
Cavendish, Henry, 21, 24
Chakras, 115, 142, 158, 160, *Plate 15*
Chaotic state, ideal, 177, 188–89, 191
Chemistry, 21–22
Chinese cultural circle, 159, 161–63
Chinese medicine
    Classical (CCM) vs. Traditional (TCM), 1
    coherence in, 194
    dynamic balance in, 151
    emotion and, 115, 194
    modern science and, 169
    popularity of, 48
    Triple Burner Meridian in, 142–43
    "unscientific" nature of, 46–47
    Western investigation of, 46
Chronic disease, 197–98, *Plates 19–21*
Classical Chinese Medicine. *See* Chinese medicine
Coherence
    coupling and, 183–86
    degrees of freedom and, 179–80
    as dynamic concept, 196–97, *Plates 17 and 18*
    homeopathy and, 129
    idealized state of, 178, 190, 191
    importance of, 172, 183
    measuring, 181, 183–92
    in medicine, 194–200
    miracle of, 180
    in music, 172–74, 180
    pyramid, 192–93, 196, 198–99, *Plate 16*
Color blindness, 138
Colors, 33
Columbus, Christopher, 87, 92, 94, 95
Combinatorics, 179
Complementary medicines
    as alternative medicine, 43–44
    chronic disease and, 198

decline of, 43–44
effectiveness of, 50
electromagnetic body and, 114–16
popularity of, 49
research on, 49–50
revival of, 43, 45–46, 50–51
scientific attitude toward, 38–39, 49
Complex systems, studying, 154–55, 182–83
Confucianism, 58, 59, 71, 161, 162
Conscience, 204–6
Consciousness
law of conservation of, 18
particle, 23
science and, 202–6
Copernicus, Nicolaus, 19, 168, 182
Copper snake, 151, 152
Coupling, 183–86
Croon, Richard, 47, 64, 76
Crystal state, ideal, 177, 178, 189–90, 191
Cultural Revolution, 208

D

Dalton, John, 21, 22, 24
Dante Alighieri, 166–67
David, King, 164
De Lavoisier, Antoine-Laurent, 21, 24
Delta distribution, 189–90, 191
Democritus, 20, 21, 23, 24–25, 26, 27, 28
Descartes, René, 54
Dirac, Paul, 25
Dissipative structures
discovery of, 94, 205
examples of, 87
standing waves as, 95–97
static vs., 87–88
ubiquity of, 87
Divine Comedy, 166–67
DNA, 22, 203
Double-slit experiment, 35, 98–99
Dowsing, 139

E

Eastern medicine. *See also* Chinese medicine
function vs. structure in, 54–55
symbol of, 151
Western doctors and, 46
ECG (electrocardiography), 38
Eckhardt, Bruno, 155
Economics, 155
Eddington, Arthur Stanley, 182
EEG (electroencephalography), 38
Einstein, Albert, 24, 26, 125, 171, 200, 201–2, 204, 205, 209–10
Electromagnetic body
challenges of studying, 110–13
chemical body vs., 109–10, 116
complementary medicines and, 114–16
complexity of, 110
dynamic nature of, 110, 111
generation of, 39
image of, *Plate 1*
Electromagnetic fields. *See also* Electromagnetic waves
discovery of, 18
ethereal qualities of, 106
image of, 19
oversensitivity to, 139
Electromagnetic waves
acoustic waves vs., 122–24
discovery of, 19
importance of, 19
movement direction of, 122–23
spectrum of, 29–30
speed of, 124
vacuum as medium for, 124–26
Electrons
anti-, 25
atomic models and, 23–24
discovery of, 23
Emotion, 115–16, 194
Emptiness, 25–26
Energy

accumulating, 120–21
dispersed vs. hard core of, 27
intangible nature of, 17
law of conservation of, 17–18, 88
matter and, 24, 26, 205
resonance and, 120–22
spirit and, 17
transferring, 121–22
Energy medicine, 38, 101
Engels, Friedrich, 208
Entropy
definition of, 182, 205
irreversible increase in, 89–90
maximum, 93
negative, 92
Equilibrium state, 93
Ernsthausen, W., 76
Ether, 124–25
Euclid, 153, 154, 168

F

Faraday, Michael, 18, 111
Far from equilibrium state, 93
Five elements, theories of, 152
Foundation for Integrated Medicine, 49
Fourier, Joseph, 186
Fourier transformation, 186
Frequency
beat, 99, 185
fundamental, 96–97
resonance and, 120–22
of standing waves, 95–97
Freud, Sigmund, 16
Fu, Jin, 172
Function vs. structure, 54–55

G

Galileo, 182
Gao, Yetao, 142–43
Gassendi, Pierre, 20
Gauss, Carl Friedrich, 188
Gaussian distribution, 188–89, 191

Gibran, Kahlil, 70
Godik, Eduard, 138
Gravitation, 17–18, 19–20, 91
Greek culture, 150, 156–57, 163, 165–66, 168, 169
Gurwitsch, Alexander, 107

H

Haken, Hermann, 52, 154
Hebrew culture, 163–66, 168, 169
Heisenberg, Werner, 20, 34, 131–32, 147, 199–200, 202, 204
Herbal medicines, 171–72
Hilbert, David, 191
Holism
coherence and, 172
history of, 194
need for, 149–50
in physics, 52
Homeopathy, 114–15, 129
Hu, Xiang-long, 76
Hu, Xu-liang, 208, 209
Huineng, 202
Hus, Jan, 167

I

I Ching, 152–53, 161, 194
Indian cultural circle, 159–60
Infinite dimensional space, 191–92
Information
definitions of, 126
resonance and, 126–28
selecting, 127–28
transferring, 126–27
Information medicine, 38
Interference. *See also* Interference patterns
constructive vs. destructive, 35, 98–99
definition of, 97
examples of, 35–36
importance of, 35
Interference patterns

acupuncture and, 102–3, 114, *Plates 11 and 12*
examples of, 100
modulation of, 100–103
stable, 36–37
structure of, 36–37
Invisible rainbow
beauty of, 34, 83
language and, 33–34
viewing, 30–33
Isolated systems, 89, 92

J

Jesus Christ, 133, 165, 167
John XXIII, Pope, 167

K

Kaptchuk, Ted, 43
Kellner, Gottfried, 56
Kepler, Johannes, 19, 182
Key-and-lock model, 105, 107
Kim, Bonghan, 56–57
Kirlian, Semyon D. and Valentina K., 140
Kirlian photography, 64–65, 140–41, 144
König, Herbert L., 28
Krieg, W. J. S., 63, 64

L

Language, limitations of, 33–34, 206–7
Lao Tzu, 161, 168, 200, 207
Lenin, Vladimir, 26, 208
Leucippus, 20, 21, 24, 27
Li, Ding-zhong, 73, 74
Li, Ke-hsueh, 52
Lie detectors, 79–80
Life
concept of, 15–16
law of conservation of, 18
quality of, 195
Liu, Zhong-Shen, 171
Log-normal distribution, 190, 191
Longo, Giuseppi, 194

Luminescence, 137–38, *Plate 14*
Luther, Martin, 167

M

Mach, Ernst, 131
Maciocia, Giovanni, 54, 67
Magnetism, 18–19
Mandel, Peter, 140
Mao Ze-dong, 48, 134, 208
Marconi, Guglielmo, 19, 105, 111
Martin V, Pope, 167
Marx, Karl, 208
Marxism, 14
Material composition, law of, 21
Materialism
in biology, 15–16
challenges to, 26
conquest and, 194–95
definition of, 14, 26, 202
Marxism and, 14
in physics, 14–15, 51
in psychology, 15, 16, 23
Matter
anti-, 25
energy and, 24, 26, 205
Maxwell, James Clerk, 18–19, 111
Medicine. *See also* Complementary medicines; Eastern medicine; Western medicine
auras and, 140–41
coherence in, 194–200
evolution of, 47, 53
present situation of, 113
science and, 44
Mediterranean cultural circle, 159, 163–69
Meridians
ancient depictions of, 59
in animals, 66, *Plate 5*
as channels, 65–66, 81, *Plate 7*
definition of, 47
movement of, 73–74
in plants, 66, *Plate 6*

shape of, 72–73
as third balance system, 67–68
width and depth of, 61
Mind
  remote communication through, 130–31
  science and, 202–3
Modulation, 126–27
Molecular biology, 22–23, 28, 108, 147
Molecules, ball-and-stick models of, 21, 22, *Plate 2*
Morphogenetic field, 16, 108–9
Morse, Samuel, 105
Motoyama, Hiroshi, 80
Music
  coherence in, 172–74, 180
  frequency distributions of, 186
  inaudible, 37–38
  order in, 173

N

Nakatani, Yoshio, 47, 64
National Center for Complementary and Alternative Medicine, 49
Nervous system, 105
Newton, Isaac, 17, 20, 182
Nile River culture, 150–51, 163
Normal distribution, 188

O

Objectivity, 201–4
Observer, influence of, 201–2, 204
Ohm's law, 81
Open systems, 92, 154–55
Overhof, C.-E., 76
Overtones, 97

P

Pair production, 25
Paul the Apostle, 131, 165
Pentateuch, 160

Perpetual motion, 88–89
Philosophical systems, 208–9
Physics. *See also* Quantum physics
  invisible and untouchable fields in, 18–20
  materialism in, 14–15, 51
  as spiritual discipline, 16–18
Pink noise, 186
Placebo effect, 132
Planck, Max, 209–10
Plato, 166
Popper, Karl, 164
Potency rule, 114
Powell, Andrew, 139
Prigogine, Ilya, 52, 69, 91, 95
Probability distributions, 187–91
Psychology
  etymology of, 15
  materialism in, 15, 16, 23
Pythagoras, 194, 200

Q

Qi, 11, 47, 58, 68–69, 135–36, 157–58
Qigong, 130, 131, 133–37
Quantum physics, 20, 22, 24, 200, 202

R

Radio waves
  discovery of, 12
  fundamental frequency of, 126
Reality, concept of, 131
Reductionism
  conquest and, 194–95
  definition of, 51
  limitations of, 51–52, 148–50, 155, 183, 195
  medicine and, 147–48
  science and, 147–48
  successes of, 52
Reformation, 167
Relativity, theory of, 125, 182, 201–2
Renaissance, 166–67

Resonance
  belief as, 132
  catastrophe, 120–21
  energy accumulation and, 120–21
  energy transfer and, 121–22
  importance of, 119
  information and, 126–28
  wireless communication and, 127–29
Rothe, H., 76
Rutherford, Ernest, 24

S

Sachs, Lothar, 190
Samsung, 138
Schlebusch, Klaus Peter, 73
Schrödinger, Erwin, 92, 131, 203–4
Science. *See also individual disciplines*
  acceptance in, 13
  basic research vs. applied, 7
  challenges of, 7, 9–10, 12
  consciousness and, 202–6
  deification of, 169
  Hebrew and Greek cultures and, 168–69
  medicine and, 44
  nature of, 209
  objectivity and, 201
  reductionism and, 147–48
  revolutions in, 52–53
Scientific method, 167, 168
Scientific spirit, 167–68, 210
Sensation propagation
  causes of, 63–64
  definition of, 57–58
  direction of, 62
  ear-needling and, 62
  enhancing, 62
  evidence of, 58
  routes of, 59–61
  speed of, 62, 103
Senses, limitations of, 28–30, 206
Shannon, Claude E., 126

Sima Qian, 164
Similarity
  principle of, 114
  statistical self-, 80
Skin resistance measurements
  as energy distribution measurements, 81–83
  as misnomer, 78–81
  reliability of, 74–76
  stability of, 77–78
Smith, Cyril W., 114–15, 129, 139
Solar wind, 32
Spirit
  energy and, 17
  law of conservation of, 18
  science and, 204–6
Stalin, Joseph, 208
Standing waves
  as dissipative structures, 95–97
  frequency of, 95–97
  one-dimensional, 36
  superposition of, 97–100
Stapp, Henry P., 14, 23, 27, 51
Stöckmann, Hans-Jürgen, 101, 155
Structure. *See also* Dissipative structures
  function vs., 54–55
  static vs. dissipative, 87–88
Subjectivity, 201–4
Succussion, 114
Synergetics, 52, 154–55
Szent-Györgyi, Albert, 15–16, 17

T

Tacoma Narrows Bridge, 120–21
Tai-chi, 151, 152, 198–99
Taoism, 161
Tao Te Ching, 161
Theory of Everything, 91
Thermodynamics, second law of, 89–90, 205
Thinking. *See also* Holism; Reductionism
  combining different ways of, 155–56

Eastern vs. Western structures for, 150–54, 156
integration of tripartite, 156–58
Third balance system hypothesis, 67–68, 83
Thomson, Joseph John, 23–24
Thomson, William, 17
Traditional Chinese Medicine. *See* Chinese medicine
Triple Burner Meridian, 142–43
Truth, 209

U
Uncertainty principle, 204
Upanishads, 160

V
Vaccines, 44
Vacuum, 24–27, 124–26
Vedas, 160
Vibrational medicine, 38
Voll, Reinhold, 47, 64, 74

W
Wave guide channel hypothesis, 68–69
Waves. *See also* Electromagnetic waves; Interference; Radio waves; Standing waves
  carrier, 126
  coupling of, 183–84
  superposition of, 34, 97–100
Wei, Mon-Zhao, 67, 83
Western medicine
  emotion and, 115
  limitations of, 45, 47, 51–52, 197
  reductionism and, 147–48, 171
  rise of, 44
  structure vs. function in, 54–55
  symbol of, 151
White noise, 186
Wiener, Norbert, 126
Wireless communication
  within the body, 105–9
  between onions, 107–8, *Plate 13*
  resonance and, 127–29
Wycliffe, John, 167

Y
Yellow River culture, 150–51
Yi, Shi, 75
Yin and yang, 151, 152, 198–99

Z
Zhang, Bi-wu, 65, 68–69, 103
Zhang, R. J., 72
Zhang, W. P., 73
Zioutas, Konstantın, 78

ABOUT THE AUTHORS

CHANGLIN ZHANG, a Chinese-trained physicist who studies the electromagnetic field in human bodies and in other living systems, has more than thirty years experience in medical research in China and Germany. He has published numerous papers in physics, biology, and medical journals in China, Germany, the UK, the U.S., and Switzerland, and he is the editor-in-chief of the *International Journal of Physiotherapy and Life Physics*. In addition to consulting with private industry medical research firms, he has been a professor of biophysics at Zhejiang University, China, and visiting professor at Siegern University, Germany.

JONATHAN HEANEY has taught physics in independent and international schools in Australia and China since 2007. He holds a bachelor's degree with honors in mechanical engineering, a bachelor's of commerce degree, and a graduate diploma in education from the University of Western Australia. Prior to teaching, Heaney worked as an engineer and as a business analyst. He has practiced Taoist qigong for twelve years.